ANC
ALIENS
ON THE
MOON

Adventures Unlimited Press

Acknowledgements

I would like to acknowledge these special souls who helped, loved and or supported me during the writing of this book: Shana Eva Paredes, Sherri Gaston, Neena Dolwani, my publisher David Hatcher Childress, Melissa Thompson, Alan Pezzuto, Sean David Morton, Denise Zak, my brother Dave and Bailey, and my little lights The Lady Aurora and Miss Fluffy-Muffy

ANCIENT ALIENS ON THE MOON

MIKE BARA

Ancient Aliens on the Moon

by Mike Bara

Copyright 2012

ISBN 13: 978-1-935487-85-2

Published by:
Adventures Unlimited Press
One Adventure Place
Kempton, Illinois 60946 USA
auphq@frontiernet.net

www.AdventuresUnlimitedPress.com

ANCIENT ALIENS ON THE MOON

Adventures Unlimited Press

Dedication

This book is dedicated to Richard C. Hoagland,
Erich von Däniken and Zecharia Sitchin.
The original "Big 3" of Ancient Alien theorists.

TABLE OF CONTENTS

INTRODUCTION

Our Moon is one mysterious place. It's easy for us to forget that, to simply go about our daily business while only occasionally looking up and noticing that gray-white celestial body in the sky. Yet even though we don't generally pay much attention to the Moon, it exerts a powerful influence over our planet, our oceans, our bodies, and perhaps even our consciousness. As I put it in my previous book *The Choice*:

> ...the Moon governs how fast the Earth spins, which is key to it being able to support human life. Without the Moon's calming influence, the Earth would spin so fast that the winds caused by the centrifugal force would most likely flatten us all like pancakes. A day would last only a few hours, making photosynthesis a dicey proposition. The Moon also regulates and agitates the Earth's protective magnetic field, which is so crucial to protecting us from high intensity solar flares and other types of deadly interstellar radiation. The Moon also just happens to be exactly the right size and exactly the right distance from the Earth to create perfect solar eclipses, a phenomenon where the disk of the Moon perfectly blots out the disk of the Sun. The diameter of the Moon also turns out to be 2,160 miles at the equator. It's not 2,161 miles, or 2,159 miles across, but exactly 2,160 miles. And those 2,160 miles also just happen to match the 2,160 year length of each astrological Age of the precession of the equinoxes. Some people think this is all just a coincidence; that the Moon just happened to be ejected from the Earth by some random planetary collision or somehow wandered into Earth's gravitational influence and stabilized in this perfect position from which it can protect us, regulate us, and even influence the female menstrual cycle.

1

I of course, do not agree that the Moon's existence and its influence on our daily lives is coincidental. In fact, it is quite obvious to me that the Moon is here by design; that some larger hand is at work in its placement in our solar system. In *The Choice*, I examined the idea that this larger, unseen hand might be God, with a capital "G." Here, in *Ancient Aliens on the Moon*, we'll examine whether that unseen hand might be "gods," with a lower case "g."

To those of you that are skeptical, I warn you that there is quite a bit more evidence for this idea than you might suspect. Not only are there those previously mentioned numerical coincidences where the Moon is concerned, there is also quite a bit of what the military calls "ground truth" as well. This ground truth has been provided over the last generation by a veritable army of reconnaissance missions, both manned and unmanned, from orbit and from direct, on-site inspection by our NASA astronauts. A lot of you, unless you've read my first book *Dark Mission–The Secret History of NASA*, will find this information new to you. Others will recognize this information for what it is, confirmation that we are either not in fact alone in this universe, or that our history is not quite what you have been told in the books they gave us in school.

The truth is, while there is a ton of evidence of Ancient Alien intervention here on Earth, there is even more evidence on the Moon (and more yet on Mars, but that's another book). The Ancient Alien theory, or the Ancient Astronaut theory as it used to be known, looks for evidence of extra-terrestrial (or supernatural) intervention in human history by examining ruins and mysteries here on Earth. The megalithic trilithons of Baalbek, the enduring mysteries of the great pyramids, the mathematical precision of Teotihuacán, the symbols of Nazca and the roadmap to human consciousness that is the Mayan ruins are all examples which the researchers have sought to decode and expose to the public. But in the course of this process, we have ignored the most obvious and perhaps the most easily confirmable evidence we have been shown. The ruins and mysteries that are revealed virtually right next door on our own celestial companion, the Moon.

This evidence consists of ruins and artifacts, buildings, instruments and vast industrial complexes in a place that according to all our conventional theories they simply cannot be – our Moon. Yet, as you will see in the images I'll show you, these structures are there, defiantly upright in a place where they should have been ground to dust eons ago by the Moon's unrelenting, incessant meteoric rain. Unless of course, they were somehow protected from that constant storm of meteors that bombard the Moon every day.

Not only will we prove that premise in the course of these pages, we will also ask some of the bigger questions it implies. All books, at their core, propose questions that they then seek to answer. The questions we need to keep in mind as we traverse these pages are these:

What did NASA find in their explorations of the solar system that they may have kept from the general public? In other words, did they find evidence of an advanced, extinct, extraterrestrial civilization on the Moon? And if they did, what was so destabilizing about them that they decided not to admit their discovery publically? The answer to those questions may surprise you, because it turns out NASA may have known all along that there were extraterrestrial ruins on the Moon, even before they went there. And I will show you that they even sought political and legal cover to justify keeping this critical ground truth from the American people, who after all paid for the missions—and from the world.

Next, we will look at just how truly ancient are these ruins on the Moon? I mean, if there were Ancient Aliens on Earth's moon eons ago, what were they doing there? Somehow, I doubt it was just a vacation spot for passing Annunaki or Gray aliens. Once you get a grasp of just how vast and extensive the alien bases on the Moon really are, it will quickly follow that they must have been engaged in *some kind* of extremely important activity there. Where they somehow intervening in the development of life on Earth? And if so, how and when? Did these visitors simply observe us from afar, or did they step in, manipulate our genetic code, and improve our bodies to the point that we became modern man? And did they dabble in other creatures' genetic codes, producing most of the modern plants, animals and livestock we know today?

Lastly, we will ask the big question; if there are ancient ruins on the Moon, what happened to the builders of them? The answer to this one may explain the answers to the first two questions. It may lead to the inevitable conclusion that the Ancient Aliens didn't simply pack up and leave; they may have been forced off the Moon, either by some conflict of unimaginable proportions, or by a natural calamity of the same dimensions. Either way, the answers to that question are bound to have created a ton of fear and trepidation inside the halls of NASA and at the highest levels of government.

But, all of those questions have to wait until we answer the first and most important question; just what the hell is on the Moon anyway?

CHAPTER ONE

EARLY OBSERVATIONS

One of the biggest problems with doing any kind of serious research about the Moon, whether it is of the mainstream variety or the more edgy stuff as in this book, is that nobody really knows much about the place. We know what we can see and have seen since the dawn of time; it always shows the same face to us (because of a condition called gravitational tidal-lock), it is made up of light stuff (highlands) and dark stuff (*maria*, or "seas" in Latin), it has a total land mass area roughly equivalent to the continents of Australia and Africa combined, and it also seems to have a profound physical effect on us here on Earth. Because of its gravitational influence, we have rising and receding tide. It both stimulates and regulates the strength of Earth's magnetic field, which protects us from harmful solar flares and radiation. And, according to rogue geologist Jim Berkland, the Moon may play a significant role in the frequency of earthquakes, especially along the Pacific Rim. Beyond that, even the female menstrual cycle is governed by it to some degree (the term "menstruation"

The Earth-Moon system, as seen by the Mars Reconnaissance Orbiter.

5

comes from the Latin mensis [month], which in turn comes from the Greek mene [moon]).[1] And while mainstream scientists deny that it has any effect on our consciousness, ask any waitress or bartender if people behave differently under a full moon and they will tell you that fights and erratic behavior accelerate noticeably when the full moon is out. Even the terms "lunatic" and "lunacy" are derived from this anecdotal observation of human behavior, and it probably also gave rise to the early werewolf myths of the Middle Ages.

The Moon also has an exceedingly uneven or "lumpy" gravitational field. While it is generally true that lunar gravity is about $1/6^{th}$ that of Earth due to its much lower mass, there are also areas mostly on the Earth-facing side where the gravitational field is much stronger. These areas generally match up with the darker Maria, or seas, and are called "mascons" (for Mass Concentrations) because the gravity of the Moon is so much stronger there. Why the Mascons exist is something of a mystery. According to established theory, a topographic valley, or depression—which the maria generally are—should have what scientists call a "negative gravitational anomaly," meaning that the gravitational field of the Moon is at least slightly weaker there. Instead, the gravitational field is much stronger, and as of the moment there is no really good explanation for this. Oh they've tried, blaming the mascons

Lunar mascon map showing mascons in the five of the largest "seas" on the Moon; Mare Imbrium, Mare Serenitatus, Mare Crisium, Mare Humorum and Mare Nectaris.

on the supposedly denser "basaltic lavas" the mare are supposedly made up of. But the vast mare sea Oceanus Procellarum is the biggest mare area on the Moon, and it has no mascon anomaly to speak of. All we can say for certain about the mascon basins is that there is something very dense beneath them, or something very powerful creating the gravitational anomalies.

Over the centuries, the Moon has been known by many names. While we call it simply "the Moon" (taken from the old English "mone" or "mōna"), ancient Germanic tribes called it "maenōn."[2] To the Greeks, it was known as "Selene" and to the Romans of course, "Luna," as in the previously mentioned lunatic. In ancient Persia, the moon was known as Metra, the world mother, and to the Aztec tribes it was Mictecacuiatl, a fearsome beast which traveled the heavens looking for victims to consume.

Contrary to popular belief, there is no "dark side" of the Moon. Because of its constant motion in orbit around the Earth and its own constant 27 day synchronized spin, at some point in the month the entire lunar surface is exposed to the light. We've also discovered in fairly recent times that the maria, or dark stuff, seems to reside pretty much only on the side that constantly faces Earth. The backside, or dark side, if you must, has almost no maria-type seas. This is an enduring mystery, at least to the mainstream scientists, that this book may yet shed some light on.

The Moon is about ¼ the Earth's diameter and about 1/81 its mass. The Moon is also the 5th largest satellite in the solar system, and the largest relative to its parent planet. There is no comparable arrangement of two such bodies anywhere in the observed universe. In fact, the Earth-Moon system is so unusual that science fiction author Isaac Asimov once referred to the arrangement as a "double planet." The Moon is also the only major satellite in the solar system that the Sun actually has a stronger hold on than the parent planet. Using something Asimov called the "Tug of War" value, he found that the Earth's gravitational pull on the Moon was less than half that of much farther away Sun:

We might look upon the Moon, then, as neither a true

satellite of the Earth nor a captured one, but as a planet in its own right, moving about the Sun in careful step with the Earth. To be sure, from within the Earth-Moon system, the simplest way of picturing the situation is to have the Moon revolve about the Earth; but if you were to draw a picture of the orbits of the Earth and Moon about the Sun exactly to scale, you would see that the Moon's orbit is everywhere concave toward the Sun. It is always "falling toward" the Sun. All the other satellites, without exception, "fall away" from the Sun through part of their orbits, caught as they are by the superior pull of their primary planets – but not the Moon.

—Isaac Asimov[3, 4]

Exactly why this would be the case is unknown, but we can use it simply to reassure ourselves that the Moon – and its arrangement with the Earth, is very, very strange.

While the differences between the Earth and Moon are significant, there are also some unaccountable similarities that give rise to the question of just where the Moon came from in the first place. There are at least four competing mainstream theories for how the Moon ended up in orbit around our planet. Those theories are not only somewhat contradictory, there is at least *some* evidence to support each of them.

Back in the day, before we actually went to the Moon, the popular idea was something called the "Lunar Fission Theory." First advanced in 1878 by George Howard Darwin, son of Charles Darwin, the Lunar Fission Theory argued that at some point in the past, the Earth had got to spinning so fast that it broke off or ejected a huge chunk of itself, which spiraled away and formed the companion body we see comfortably orbiting us today. The theory was that this chunk of the Earth's mantle (the thick rocky layer just below the Earth's crust, about 20-30 miles down) was somehow broken away in a violent episode, possibly caused by the Sun's gravitational pull. In other words, sometime after the Earth cooled and assumed a solid form, a chunk of the Earth's solid mass was somehow weakened, and then the sun pulled it off into a stable orbit about 240,000 miles out. The most popular location for all this material to have come from was the Pacific basin, a huge depression in the planet that at some places is as much as 35,000 feet deep and is wide enough to easily contain the entire continent of Africa. Since the Moon is about 1/4th the size of the Earth, the Pacific basin could easily be the source of the rocky material necessary to create the Moon itself.

The Lunar Fission Theory would also neatly explain some of the enduring mysteries of the Moon. While the Moon is 1/4th the size of the Earth, its gravity is, as I mentioned, only 1/6th that of Earth. The discrepancy is accounted for by the fact that Moon has a much lower overall density and less mass. Being made up mostly of lighter materials normally found in the Earth's crust or mantle (the layer right beneath the crust) the Moon also has a relatively tiny core. Only about 3 percent of the Moon is thought to be made up of a heavier nickel-iron core, whereas the Earth's nickel-iron core makes up about 30 percent of its much greater mass.

If in fact the Moon was somehow spun off from the Earth itself, all of this would make sense. If the material required to make the Moon came from the lighter upper mantle of the Earth, then the Moon would obviously lack the heavier core elements that the Earth possesses. As the material that broke away collapsed and

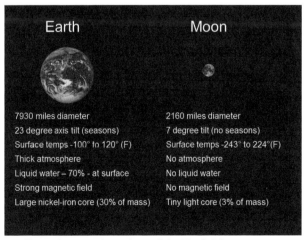

General comparison chart of the Earth-Moon system.

compressed, these heavier elements would sink to the middle and form the core. In fact, such an exotic composition is exactly what we would expect if the Lunar Fission Theory were substantially correct. However, there are three mechanical objections to the Lunar Fission Theory that would seem, on the surface anyway, to discredit it, at least to some degree.

The first objection to the fission theory is that the Earth-Moon system simply lacks the required *angular momentum*, or spin energy, for it to work. In order for a solid chunk of the Earth to break away in the equatorial region and spin off into space, the Earth-Moon system would have to have about twice as much of the spin energy it currently possesses. In order for the Earth to become unstable (and the resonant vibrations necessary to achieve fission to occur), a single day would have to have been about 3 hours long, rather than the current 24. Since angular momentum is assumed to be a constant, where did all this "missing" spin energy go?

The second objection is that such an eruption would most likely be from the equatorial region of the Earth. If this were the case, then logically the Moon would orbit around the Earth's equatorial plane, much like the planets orbit around the Sun's plane of the ecliptic. Instead, we see that Moon's orbital plane is tilted 28.5 degrees to the Earth's equator.

The third objection was that the newly broken away Moon would have had devastating tidal effects on the Earth, and possibly been broken apart as it passed the Earth's destructive "Roche limit." It is argued that no evidence of such tremendous tidal disruptions exists in the geologic record today.

Despite all this, the Lunar Fission Theory remained popular well into the twentieth century. One fanciful account from a 1936 U.S. Office of Education script for a children's radio program told the story this way:

> "FRIENDLY GUIDE: Have you heard that the moon
> once occupied the space now filled by the Pacific Ocean?
> Once upon a time—a billion or so years ago—when the

Earth was still young—a remarkable romance developed between the Earth and the sun—according to some of our ablest scientists . . . In those days the Earth was a spirited maiden who danced about the princely sun—was charmed by him—yielded to his attraction, and became his bride . . . The sun's attraction raised great tides upon the Earth's surface . . . the huge crest of a bulge broke away with such momentum that it could not return to the body of mother Earth. And this is the way the moon was born!

GIRL: How exciting!" [4]

However exciting, the Lunar Fission Theory began to get some competition by the early 20th century. In 1909, an astronomer by the name of Thomas Jefferson Jackson See proposed a new idea; that the Moon had just been wandering by and was somehow "captured" by the Earth's gravitational pull and settled into a stable orbit. This scenario, while possible, is highly improbable for a number of reasons. First, celestial objects tend to move through the vacuum of space pretty quickly. The Earth, for instance, travels at about 67,108 mph through space, which generates quite a bit of inertia, or momentum. How the Earth's relatively weak gravitational field could capture another object moving past and pull it into a stable orbit is a problematic question with no easy answer. In an attempt to resolve it, the capture theory was modified so that the Earth in the distant past had a much denser atmosphere that was also much greater in volume. If this had been the case, the dense atmosphere could have helped slow down the wandering Moon, but so far no evidence supporting this proposal has ever emerged.

And there were other problems. Even without the highly expanded atmosphere idea, the intricate celestial dance required to make the capture theory work would be incredibly complex and almost unimaginably coincidental. Since nothing like it has ever been observed anywhere in the universe, it remained a very unlikely possibility even before the astronauts landed on the Moon and brought back rock samples. It was really then that the capture theory completely fell apart.

Logically, if the Moon was just wandering by and somehow magically captured by the Earth's gravitational field – which remember is weaker than the Sun's hold on it—then a couple of assumptions therefore follow. The first is that since the Moon by definition would have formed somewhere else, it should not be made up of materials similar to Earth or of the same relative age. That's where the moon rocks came in. What they showed is that not only is the Moon made up of the same "stuff" as the Earth, it was, like the Earth, formed some 4.5 billion years ago. So all things considered, the capture theory didn't ever really get off the ground. But it did give rise to another idea, which became all the rage for a period of time in the late 1970's. This was the "co-accretion theory."

The co-accretion theory arose from the accretion theory of planetary formation (which I thoroughly dismantled in my last book, *The Choice*). This idea, initially advocated by the French astronomer Edouard Roche, argues that planets are formed by simply coalescing from the leftover dust and debris of exploded stars. These debris clouds are called "nebula" by the astronomical community, and the idea is that clumps of material begin to form in these primordial nebula, run into each other, magically glue themselves together, and eventually become planets. Roche simply expanded the notion so that the Earth and Moon formed literally side by side, just as we see them today. However, the co-accretion theory cannot account for why the Moon is so much less dense than the

Earth, or why it has such a small core and virtually no heavy elements. Logically, if they formed together in the same region of the primordial soup, they should have similar compositions and densities. Not only that, but the co-accretion theory

Artists depiction of the capture theory. (NASA)

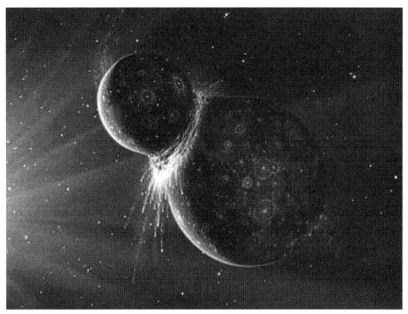

could not account for the high amount of angular momentum (spin energy) in the Earth-Moon system.

So, having pretty much struck out by the late 1970's, the planetary scientists began to seize on a new theory of how the Earth-Moon system formed; the "Big Whack" theory.

Re-dubbed the Giant Impact Hypothesis, this new idea said that a large, Mars sized object struck the Earth sometime in the distant past (about 4.5 billion years ago, by most estimates). This impact ejected a huge amount of material off the Earth and into the Moon's orbit, where it cooled, coalesced and formed a completely new body we know today as our sister Moon. This massive object is sometimes called *Thea*, after the Greek Titan that was the mother of Selene, the Moon goddess of ancient Greek mythology. Advocates of the Giant Impact Hypothesis (let's just call it the GIH from now on) point to what they argue are several lines of evidence to support it. Primary among these are that the Earth's and the Moon's orbit are in the same direction (possibly indicating they had a common origin point in the solar system), and the fact (derived from moon rocks brought back the Apollo astronauts) that the lunar surface was once nearly entirely molten.

Of course, there are also problems with the GIH. First, an impact such as the one required to make the GIH work would have probably melted the entire surface of the Earth into a molten magma ocean, at least for a brief period of time. Since there is no geologic evidence that such a global melt ever took place, that finding in of itself pretty much shoots down the GIH theory. Further proof that the GIH is wrong was recently found in the study of titanium oxygen isotopes from the Moon, Earth and meteors. The study, which is kind of a planetary paternity test, found that Moon rocks and Earth rocks had virtually identical oxygen isotopic ratios, meaning that the Earth and Moon are chemically exactly the same. This has two implications; first that the Moon formed *from* the Earth, and second that there was no giant impacting body that struck the Earth and formed the Moon. If there was, the oxygen isotopic ratios of the Earth and Moon would be different, since it is virtually guaranteed that an object which formed elsewhere in the solar system would not have identical ratios to the Earth's. In other words, if Thea had truly struck Earth, Thea and Earth's chemicals and elements would have mixed when the Moon formed, and the Moon would be a compositional mixture of the two. It is not. It is exactly like the Earth, meaning it either formed near the Earth, or was somehow broken away from it.

So much for Thea and the GIH.

Unfortunately, this leaves us without a single viable working theory for how the Moon formed. Or at least, it leaves the planetary geologists and astronomers without one. Fortunately, just as I did in my previous book *The Choice*, I'm here to sort things out for the planetary geologists and give them a new and more likely theory they can hang their hat on. It's called the Solar Fission Theory.

The Solar Fission Theory, chiefly advocated by the late Dr. Tom van Flandern in his book *Dark Matter, Missing Planets and New Comets: Paradoxes Resolved, Origins Illuminated*, argues that the Sun spins off the planets from its belly very early in its life-cycle, and that the planets subsequently give birth to their moons in the same manner. As I described it in *The Choice*:

"In the solar fission model, once the biggest chunk of the solar

14

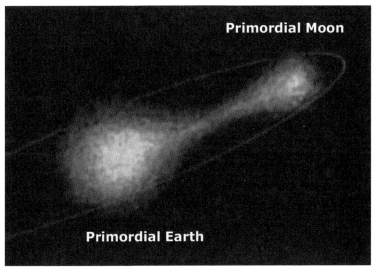

nebula collapses and begins nuclear fusion (ignition into a star), it starts sucking up all the nearby dust and gas and it quickly grows in size. By adding all this fuel to its nuclear furnace, it soon begins spinning so fast that the centrifugal forces become stronger than the gravitational field of the newborn star. At this point, the star "oblates," or bulges at the center, and solar material is flung out from the equatorial region of the young star. This material then spirals outward in (roughly) twin pairs, forming first gas giant planets and then later "terrestrial" or rocky planets like our Earth. As the star gives birth to pair after pair of twins in this manner, its angular momentum (spin energy) dissipates and the star begins to emit energy in a stable cycle ideal for supporting life giving planets. As the blobs of ejected "star stuff" spiral away from their birth mother, they also give birth in turn to their own moons, and eventually find their resonant orbits and begin to cool. After a billion years or so, the whole system should achieve a state of equilibrium and balance. The planets will cool. On some of them in the habitable zone, like Venus, Earth or Mars, water will form oceans, bacteria will start the cycle of life, and the children of this elegant birthing process will eventually walk the face of these planets, stare into the night sky, and wonder how they got there in the first place."

The theory then goes on to explain that large gas giant planets

15

like Jupiter and Saturn will tend to spin off multiple moons (in roughly twin pairs) and smaller terrestrial planets like Earth and Venus will tend to spin off only one large moon, much like we see in the Earth-Moon system. In van Flandern's model, Earth and the Moon are an example of such a pairing, and so are Venus and Mercury, with Mercury having been ejected from its orbit around Venus by some ancient impact.

The chief objection to the idea that the Moon broke off from the Earth was the fact that it would take far more spin energy than the current Earth-Moon system possess to actually break a chunk of solid Earth off. The Solar Fission model also solves this problem. In van Flandern's model, the Moon didn't break away from the primordial Earth *after* it cooled and solidified, it spun off out of the early *molten* Earth. This would also explain why the Moon is made up primarily of material from the Earth's lighter mantle, rather than the heavier iron-rich core. The only observation that isn't accounted for is the fact that the Moon's orbital plane is inclined by 5.14° to the Earth's. However, there could be numerous explanations for this (like later impacts which forced the Moon to a different position) and so this is not a show-stopper for the theory.

What does appear to be certain is this; whatever the Moon's origins, it appears to be either formed from the Earth itself or very nearby at the same time as the Earth, 4.5 billion years ago. Fanciful stories of it being a celestial body from another part of the solar system (or the galaxy) that was "driven" here and placed into orbit are most likely wrong. That does not however preclude it from being inhabited or modified much later in its evolutionary process. For instance, sometime after life began on Earth.

Early Studies and Anomalies

Almost from the dawn of human history, the ancients studied the Moon and began to unravel its secrets. The ancient Greek philosopher Anaxagoras (4th century BC) speculated that the Sun and Moon were both giant spherical rocks, and although he got that wrong, he did correctly surmise that the Moon was visible because

16

it was reflecting the light of the Sun. The Chinese astronomer Shi Shen published instructions for the predictions of both lunar and solar eclipses in the 4th century. Jing Fang (78–37 BC), of the Chinese Han dynasty, also correctly predicted that the Moon was spherical. In the 2nd century BC, Seleucus of Seleucia correctly theorized that tides were due to the attraction of the Moon, and that their height depended on the Moon's position relative to the Sun. Ptolemy of Egypt (90–168 AD) calculated the distance of the Earth to the Moon and its size relative to the Earth with amazing accuracy. He calculated the Moon was at a distance of 59 times the Earth's radius and had a diameter of 0.292 Earth diameters. The actual values are 60 and 0.273 Earth diameters. Babylonian astronomers were the first to record the 18-year cycle of lunar eclipses in the 5th century AD. In 499 AD, the Indian astronomer Aryabhata stated in his journal the *Aryabhatiya* that reflected sunlight is what causes the Moon to shine.

That all of this was discovered or accurately predicted before the invention the telescope is somewhat astonishing. By the

Transient Lunar Phenomena (TLP) visible on the Moon. (NASA)

time Galileo Galilei made his first drawings of the Moon from his telescopic observations in 1609, most of the world had come to believe that while the Moon was spherical, its surface was probably glass smooth. Galileo was the first to truly insist that it was in fact made up of mountains and valleys – much like the Earth – and that the deep craters visible on its pockmarked surface were probably the results of volcanic activity (which some are) or massive impacts.

Through this same time period, observers began to note what appeared to be temporary changes in the surface appearance of the Moon. These "Transient Lunar Phenomena" (TLP's or LTP's for short) are usually noted as observations of changes in the color or reflectivity of the lunar surface. Since the 1600's, observers have cataloged at least 579 separate observations of TLP's, according to one NASA report.[5] Most of these events last anywhere from a few minutes to a few hours (hence the "transient" part of the designation) and have a tendency to cluster around certain areas of the lunar surface. The craters Alphonsus and Aristarchus are the two most commonly mentioned in the literature.

Descriptions of TLP's range from foggy patches to what appear to be mist-like or cloudy formations or other forms of obscuration of the lunar surface. Changes in coloration ranging from red, green, blue or violet have been noted, as have areas of increased brightness or areas of increased darkness. Two extensive catalogs of TLP's exist, with the most recent cataloging some 2,254 events going back to the 6th century. Of those that are considered the most reliable reports, about 1/3 come from the vicinity of the previously mentioned Aristarchus plateau.

Wikipedia lists some of the more well-known examples of TLP reports as follows:

•On June 18, 1178, five or more monks from Canterbury reported an upheaval on the moon shortly after sunset. "There was a bright new moon, and as usual in that phase its horns were tilted toward the east; and suddenly the upper horn split in two. From the midpoint of this division a flaming torch sprang up, spewing out, over a considerable distance, fire, hot coals, and sparks. Meanwhile the body of the moon which was below writhed, as it were, in anxiety, and, to put it in the words of those who reported it to me and saw it with their own eyes, the moon throbbed like a wounded snake. Afterwards it resumed its proper state. This phenomenon was repeated a dozen times or more, the flame assuming various twisting shapes at random and then returning to normal. Then after these transformations the moon from horn to horn, that is along its whole length, took on a blackish appearance." In 1976, Jack Hartung proposed that this described the formation of the Giordano Bruno crater.

•During the night of April 19, 1787, the famous British astronomer Sir William Herschel noticed three red glowing spots on the dark part of the moon. He informed King George III and other astronomers of his observations. Herschel attributed the phenomena to erupting volcanoes and perceived the luminosity of the brightest of the three as greater than the brightness of a comet that had been discovered on April 10. His observations were made while an aurora borealis (northern lights) rippled above Padua, Italy. Aurora activity that far south from the Arctic Circle was very rare. Padua's display and Herschel's observations had happened a few days before the sunspot number had peaked in May 1787.

•In 1866, the experienced lunar observer and mapmaker J. F. Julius Schmidt made the claim that Linné crater had changed its appearance. Based on drawings made earlier by J. H. Schröter, as well as personal observations and drawings made between 1841 and 1843, he stated that the crater "at the time of oblique illumination cannot at all be seen" (his emphasis), whereas at high illumination, it was visible as a bright spot. Based on repeat observations,

19

he further stated that "Linné can never be seen under any illumination as a crater of the normal type" and that "a local change has taken place." Today, Linné is visible as a normal young impact crater with a diameter of about 1.5 miles (2.4 km).

•On November 2, 1958, the Russian astronomer Nikolai A. Kozyrev (prominently mentioned in *The Choice*) observed an apparent half-hour "eruption" that took place on the central peak of Alphonsus crater using a 48-inch (122-cm) reflector telescope equipped with a spectrometer. During this time, the obtained spectra showed evidence for bright gaseous emission bands due to the molecules C2 and C3. While exposing his second spectrogram, he noticed "a marked increase in the brightness of the central region and an unusual white colour." Then, "all of a sudden the brightness started to decrease" and the resulting spectrum was normal.

•On October 29, 1963, two Aeronautical Chart and Information Center cartographers, James A. Greenacre and Edward Barr, at the Lowell Observatory, Flagstaff, Arizona, manually recorded very bright red, orange, and pink colour phenomena on the southwest side of Cobra Head; a hill southeast of the lunar valley Vallis Schröteri; and the southwest interior rim of the Aristarchus crater. This event sparked a major change in attitude towards TLP reports. According to Willy Ley: "The first reaction in professional circles was, naturally, surprise, and hard on the heels of the surprise there followed an apologetic attitude, the apologies being directed at a long-dead great astronomer, Sir William Herschel." A notation by Winifred Sawtell Cameron states (1978, Event Serial No. 778): "This and their November observations started the modern interest and observing the Moon." The credibility of their findings stemmed from Greenacre's exemplary reputation as an impeccable cartographer. It is interesting to note that this monumental change in attitude had been caused by the reputations of map makers and not by the acquisition of photographic evidence.

•On the night of November 1–2, 1963, a few days after Greenacre's event, at the Observatoire du Pic-du-Midi in the French Pyrenees, Zdenek Kopal and Thomas Rackham made the first photographs of a "wide area lunar luminescence." His article in Scientific American transformed it into one of the most widely publicized TLP events. Kopal, like others, had argued that Solar Energetic Particles could be the cause of such a phenomenon.

•During the Apollo 11 mission Houston radioed to Apollo 11: "We've got an observation you can make if you have some time up there. There's been some lunar transient events reported in the vicinity of Aristarchus." Astronomers in Bochum, West Germany, had observed a bright glow on the lunar surface—the same sort of eerie luminescence that has intrigued moon watchers for centuries. The report was passed on to Houston and thence to the astronauts. Almost immediately, Armstrong reported back, "Hey, Houston, I'm looking north up toward Aristarchus now, and there's an area that is considerably more illuminated than the surrounding area. It seems to have a slight amount of fluorescence."

•In 1992, Audouin Dollfus of the Observatoire de Paris reported anomalous features on the floor of Langrenus crater using a one-meter (3.2-foot) telescope. While observations on the night of December 29, 1992, were normal, unusually high albedo and polarization features were recorded the following night that did not change in appearance over the six minutes of data collection. Observations three days later showed a similar, but smaller, anomaly in the same vicinity. While the viewing conditions for this region were close to specular, it was argued that the amplitude of the observations were not consistent with a specular reflection of sunlight. The favored hypothesis was that this was the consequence of light scattering from clouds of airborne particles resulting from a release of gas. The fractured floor of this crater was cited as a possible source of the gas.

While the most common explanation for these transient events is some sort of volcanic "out-gassing," it must be noted that this conflicts with the idea that Moon is geologically inactive. Without tectonic plate movement or an active, rotating core, the Moon could not possibly be geologically active, meaning that these mysterious coloration changes and clouds could not possibly be caused by erupting gasses. Any volatiles would have long since escaped over the Moon's 4.5 billion year life cycle. So if that isn't the explanation, then what is? A good question. Which we are about to answer…

Footnotes:

[1]*The Reluctant Hypothesis: A History of Discourse Surrounding the Lunar Phase Method of Regulating Conception*. Lacuna Press. p. 239. ISBN 978-0951097427.

[2] Barnhart, Robert K. (1995). *The Barnhart Concise Dictionary of Etymology. USA*: Harper Collins. p. 487. ISBN 0-06-270084-7.Allen, Kevin (2007).

[3]Asimov, Isaac (1975). "Just Mooning Around", collected in *Of Time and Space, and Other Things*. Avon. Formula derived on p. 89 of book. p. 55 of .pdf file. Retrieved 07-24-2012.

[4]Aslaksen, Helmer (2010). "The Orbit of the Moon around the Sun is Convex!". National University of Singapore: Department of Mathematics. Retrieved 07-24-2012.

[4] http://www.pbs.org/wgbh/nova/tothemoon/origins.html

[5]Barbara Middlehurst, Jaylee Burley, Patrick Moore, and Barbara Welther (1968). "Chronological Catalog of Reported Lunar Events," NASA TR R-277

CHAPTER 2

THE 20ᵀᴴ CENTURY

*I'm sure you're aware of the extremely grave potential for
cultural shock and social disorientation contained in the present
situation, if the facts were prematurely and suddenly made public
without adequate preparation and conditioning. Anyway, this is the
view of the Council... there must be adequate time for a full study
to be made of the situation before any thought can be given to
making a public announcement.*
–Dr. Heywood Floyd, *2001: A Space Odyssey*

The modern era of lunar exploration actually began in the late
1950's with the formation of the National Aeronautics and Space
Administration (NASA) in the United States and advent of the
modern "space race" with the Soviet Union. At the outset, no one
really knew where the space race was headed, or even where the
finish line was. All that either side knew was that space was next
great frontier to be conquered.

With the development of early rockets like the German V-2
and the U.S. Redstone, it became possible to consider launching
artificial satellites into Earth orbit. Thanks to many technological
innovations of the period, both solid and liquid fuel rockets were
under development in the West by various rocket teams. It seemed
only a matter of time until the United States placed a satellite (or a
man) into Earth orbit.

No one knew the full extent of the secretive Russian rocketry
progress until October 4ᵗʰ, 1957, when the Soviet Union stunned the
world by successfully launching the world's first artificial satellite,
Sputnik 1, into orbit. This achievement electrified the American
public, which had assumed they were well ahead in the newly
declared "space race."

Because of the possible threat of the communists raining nuclear
bombs from space with no warning, the United States responded by

hurriedly trying to launch a response mission, Vanguard 1. It blew up on the launching pad. Desperate, the U.S. turned to Dr. Werner von Braun and his handpicked team of Nazi rocket scientists to find a way to match the Soviets. Von Braun was considered a last resort because of his Nazi past, but he was a genius in rocketry, and he saved the day by successfully putting Explorer I, a 30-pound payload, into orbit.

But the euphoria of that achievement didn't last long. The Soviets launched one successful mission after another and were clearly well ahead of the U.S. in their ability to place payloads into Earth orbit. And soon they had their eyes on another prize; being first to the Moon.

The Brookings Report

In order to counter this, the United States decided that it needed a centralized space planning agency that would oversee the civilian U.S. Space efforts. NASA, as we know it today, actually evolved from several earlier organizations. One called the National Advisory Committee for Aeronautics, or NACA, was the primary source of early NASA brainpower. The NACA director, Dr. Vannevar Bush, was an instrumental player in many early aerospace projects and companies. He was co-founder of Raytheon systems, still a major defense contractor, and was director of the Office of Scientific Research and Development, which oversaw the Manhattan Project which developed the U.S. atomic bomb. He was also President Roosevelt's scientific advisor and played a key role in bringing many of the Nazi rocket scientists, like Werner Von Braun, to the USA.

But from the beginning, NASA was born under a cloak of secrecy that seemed aimed at one specific secret agenda; hiding the evidence of an Ancient Alien presence in the solar system.

Publically, the *National Aeronautics and Space Act of 1958* states that NASA is "a civilian agency exercising control over aeronautical and space activities sponsored by the United States." But in reality, the Space Act shows that NASA is a sub-division of the Department of Defense and is subject to DOD oversight of all its activities:

"Sec. 305... (i) The [National Aeronautics and Space] Administration shall be considered a defense agency of the United States for the purpose of Chapter 17, Title 35 of the United States Code..."

The Act also makes it clear that NASA is *not* free to disclose anything that the President or the DOD might consider to be "classified information."

"Sec. 205... (d) No [NASA] information which has been classified for reasons of national security shall be included in any report made under this section [of the act]..."

The intent of these sections is to make it clear that NASA is not an independent, civilian space research agency, but rather an adjunct of the DOD that is totally under that department's control. The reasons for this are not immediately clear until you study another document also commissioned at the dawn of the space age; the so-called "Brookings Report."

After NASA was formed and almost before the ink was dry on the bill that brought it into being, NASA commissioned a formal study into the projected effects on American society of its many planned activities. This report was first published to some fanfare in early 1960's, but then lay dormant for many years afterwards until it was rediscovered in the mid-1990s.

At that time, Professor Stanley V. McDaniel was seeking additional documentation for his then-ongoing study into NASA's new imaging and data policy surrounding the controversial Mars Observer program. In the final stages of his study, McDaniel asked Richard C. Hoagland (my co-author on *Dark Mission*) for some assistance in locating some NASA documents and research papers relating to its SETI (Search for Extraterrestrial Intelligence) project. Hoagland told McDaniel of the long-rumored existence of an official NASA report supposedly commissioned by the space agency in its early years that related to possible NASA censorship of SETI evidence if it was ever discovered. At McDaniel's urging, Hoagland began actively searching for the document, polling various contacts and eventually having a conversation with former police detective Don Ecker. Ecker, a consultant to *UFO* magazine, called in a couple of favors and not only confirmed the existence of this highly controversial study—but came up with the actual title: *Proposed Studies on the Implications of Peaceful Space Activities for Human Affairs*.

After some more digging, Hoagland eventually came up with the document, authored by The Brookings Institution. The Brookings Institution was probably the world's foremost "think-tank" of its day, and the contributors to the NASA study were a veritable "who's-who" of the leading academics of the time. MIT's Curtis H. Barker, NASA's own Jack C. Oppenheimer, and famed anthropologist Margaret Mead were all consulted for contributions to the final Report.

After scouring the document, it quickly becomes apparent that the underlying purpose of the Brookings Report was to provide legal and political cover for NASA in the event it ever kept secret discoveries that the president or the DOD declared "classified." The most stunning remarks came on page 215, where the Report mentions the possibility that "artifacts" (i.e. extraterrestrial Ancient Alien ruins) may be found by NASA in their explorations of the solar system:

"While face-to-face meetings with it [extraterrestrial intelligence] will not occur within the next twenty years (unless its technology is more advanced than ours, qualifying it to visit Earth), *artifacts* **left at some point in time by these life forms might possibly be discovered through our space activities on the Moon, Mars or Venus"** the Report states. "Artifacts" of course, mean ruins. Ancient Alien ruins. It is obvious from this statement that NASA not only *suspected* they might encounter ancient alien ruins, they *expected* to find them.

It then goes on to consider whether such a discovery should, rather than be made public immediately, be *suppressed*:

"How might such information, under what circumstances, be presented or *withheld* from the public, [and] for what ends?"

This single line alone in the 250 page Report provides NASA

with the crucial political cover it needed to justify hiding any discoveries of Ancient Alien ruins anywhere in the solar system. It justifies the possible cover-up by considering the possibility of social devastation if such a discovery were made public without an adequate preparation period for social adjustment:

"Anthropological files contain many examples of societies, sure of their place in the universe, which have *disintegrated* when they had to associate with previously unfamiliar societies espousing different ideas and different life ways: others that survived such an experience usually did so by paying the price of changes in values and attitudes and behavior..."

The Report then goes on to reinforce the point with more examples:

"...the fundamentalist (and anti-science) sects are growing apace around the world... For them, the discovery of other life — rather than any other space product—would be electrifying... If super-intelligence is discovered, the [social] results become quite unpredictable..."

Obviously, this section is particularly relevant even today, with the rise of Islamic fundamentalism all over the world. But the Report also cautions that the more advanced and scientifically educated nations and individuals, such as those in the United States, might also suffer similar emotional and social upheaval:

"... of all groups, scientists and engineers might be the most devastated by the discovery of relatively superior creatures, since these professions are most clearly associated with mastery of nature." (p. 225)

It then suggested, obviously, that further studies were needed. It then goes on to make the following (and somewhat alarming) statement about the entire question of whether to announce or withhold the discovery of Ancient Alien ruins:

"... the consequences of such a discovery are presently unpredictable..."

The Report then references an obscure work by psychologist Hadley Cantrell, titled *The Invasion From Mars: A Study in the*

Psychology of Panic (Princeton University Press, 1940). This little known book was commissioned by the Rockefeller Foundation under a grant to Princeton University. Its subject was the 1938 Orson Welles *War of the Worlds* radio broadcast and the effect it had on portions of the American listening public. It is estimated that more than a million people in the northeast United States panicked over the broadcast, hearing Wells' brilliant production and believing the Martian invasion was real. The implication of the book and its inclusion in the Brookings Report is that the broadcast was a psychological warfare experiment, and that America dramatically failed the test. Given this, it seems reasonable that NASA might have adopted an official policy of cover-up of the discovery of any Ancient Alien artifacts it might discover.

It isn't difficult to sum up the Brookings Report. Among its wide-ranging analysis and conclusions are the following:

1. "Artifacts" (i.e. Ancient Alien ruins) are likely to be found by NASA on the Moon and\or Mars.

2. If the artifacts point to the existence of a superior civilization, the social impact is "unpredictable."

3. Various negative social consequences, from "devastation" of the scientists and engineers, to an "electrifying" rise in religious fundamentalism, to the complete "disintegration" of society are distinct possibilities. The War of the Worlds broadcast provides an excellent example.

4. Serious consideration should be given to "withholding" such information from the public if, in fact, artifacts are ever discovered.

So here we have the proverbial smoking gun. Not only was NASA advised almost from its inception to withhold any data that supported the reality of the Ancient Alien theory or any other discovery like it, they were told to do so for the good of human society as a whole. Most especially, they should withhold the data from their own rank and file engineers and scientists, since they were the most vulnerable members of all of human society. It doesn't take a proverbial rocket scientist to conclude that NASA took these recommendations and transformed them into policy at the highest levels. Nor would it be surprising if the whole question of "artifacts" were considered a national security issue – given (again) NASA's founding charter position as "a defense agency of the United States."

Although the document itself is fairly obscure today, this was not so in the early 1960's. The New York Times published a brief summary of the Brookings Report in December, 1960. It was apparent from the Times' treatment of the Report that the potential for a social meltdown if such explosive information ever became public was considered a prime threat to the existing social order. "MANKIND IS WARNED TO PREPARE FOR DISCOVERY OF LIFE IN SPACE: Brookings Institution Report Says Earth's Civilization Might Topple if Faced by a Race of Superior Beings," the December 15th, 1960 headline screamed breathlessly.

As part of our research for *Dark Mission*, Richard C. Hoagland also discovered that The Brookings Report was the basis for Arthur C. Clarke and Stanley Kubrick's seminal film *2001: A Space Odyssey*. In fact, according to a 1968 *Playboy* interview, Kubrick could quote from the Brookings Report chapter and verse. In the interview, he quoted the exact passages shown above, and declared that the whole question of covering up the discovery of artifacts to be the central theme of his legendary film.

From early on, Brookings officially affirmed NASA's expectation that the agency would inevitably fly to nearby planets in the solar system, and would thus be physically capable, for the first time, of confronting "extraterrestrials" right in their own backyard. Obviously, this goes a long way to explaining the sometimes irrational "skepticism" that most mainstream NASA- funded scientists have regarding the whole ET question. It also might go a long way to explaining some of NASA's later behavior as their exploration of the solar system actually began.

Early Unmanned Probes

Armed with this new legal and political cover, both NASA and the Soviet space programs were free to begin going to the Moon and beyond. The Russian were the first to try.

The Soviet Luna program began in 1957 and its objective from the beginning was to successfully send an unmanned probe to the Moon and crash it there. While this may seem like a goal of questionable value today, at the time it would have been a major achievement on the scale of Sputnik itself. But when the idea was first conceived, no one knew for certain just how difficult it would turn out to be to actually navigate to the Moon, much less to the Moon and back as would be necessary for a manned mission to our nearest neighbor.

As I extensively documented in *The Choice*, and as Richard C. Hoagland and I covered in the extended edition of *Dark Mission*, it turned out that simply launching a probe into Earth orbit and then aiming it at and actually hitting the Moon was quite a challenge. Flatly, it should not have been. The mathematics involved and the calculations for gravity were well known, even in the late 1950's. All that should have been required was to get a probe up into space and then fire the rocket to the point the Moon would be in a few days. The Moon, after all, has a diameter of more than 2,160 miles. That's a pretty big target. But neither the Russians nor the Americans could seem to figure out this seemingly straightforward task.

The Russians were the first to try. Their first three attempts, named Luna-1958A, 1958B and 1958C all failed in the ascent stage of the mission due to problems with the boosters. It wasn't until Luna 1, actually the 4th Luna mission, that the Soviets finally got a Luna probe into orbit (the Russians had a habit of not officially recognizing their unsuccessful missions). Once there, the Russians took aim, fired their 3rd stage rocket, and propelled Luna 1 at its intended target – the Moon, some 239,000 miles away.

And they missed. By far more than the proverbial mile. By 3,725 of them to be exact.

Again, not to overemphasize points I have already made in *The Choice* and *Dark Mission*, but that is simply not possible if Newtonian mechanics is correct. To miss the Moon by more than one and a half times its own diameter when you're already weightless and in

Earth orbit is pretty much impossible, unless the laws of physics are somehow far different than we have been led to believe.

Meanwhile, the U.S. didn't fare much better. In early 1959 a JPL constructed satellite, Pioneer 4, missed the Moon by a whopping 37,000 miles, more than *17 times* the Moon's diameter! Obviously, there were issues with navigating the space between Earth and our nearest neighbor, issues which were not successfully solved until many years later, when the Ranger 4 spacecraft lost all power shortly after achieving orbit, and subsequently did what no other American probe had yet been able to do – actually *hit* the Moon. Eventually, von Braun and NASA figured out that having active, spinning systems on board the spacecraft added energy to the system, and once they empirically accounted for this the era of modern lunar exploration really began (see *The Choice*).

The U.S. programs started with the Ranger series in the early 1960's. These were designed to fly into the Moon at high velocity, transmitting back television images the entire way to give humans the first close-up views of the lunar surface. As I discussed in *The Choice*, there really wasn't a fully successful Ranger mission until Ranger 7 in 1964, but the program allowed von Braun and NASA to figure out the hidden physics of interplanetary navigation. Ranger was soon followed by two new and parallel programs, Lunar Orbiter, which was to take reconnaissance photographs and map the entire surface of the Moon, and Surveyor, which was to test the ability to soft land on the lunar surface and study it.

The Lunar Orbiter series was highly successful, with all five missions being completed essentially as planned. Lunar Orbiter 1 was obviously the first in the series, launched in August 1966, and it was to photograph the equatorial section of the Moon's near side so NASA could scout for possible manned landing sites. From an altitude that varied from 36 to 25 miles above the lunar surface, Lunar Orbiter 1 took some 42 high resolution and 187 medium resolution "frames" of the lunar surface (the camera was a line-by-line scanning system that transmitted images back to Earth as "framelets" that were lined up and reassembled as full images on Earth). The most famous image from Lunar Orbiter 1 was of the first view ever of the Earth from lunar orbit.

Once its mission was complete and its film exhausted, the spacecraft was deliberately crashed into the lunar surface in order to test NASA's ability to remotely track the vehicle.

Like Lunar Orbiter1, Lunar Orbiter II (November, 1966) was designed as a landing site reconnaissance mission of the Moon's equatorial region from high altitude (32 miles). It took a total of 609 high resolution and 208 medium resolution frames but because of the altitude, they were of questionable value. Lunar Orbiter II did become famous for an oblique view of the peak of the crater Copernicus which was hailed at the time as one of the great photos of the 20th century. But the most interesting photo Lunar Orbiter II took was by far image number LO2-61H3, which showed peaks that came to be known simply as "The Blair Cuspids."

On November 22, 1966, – three years to the day from the date President Kennedy had been killed -- NASA released a Lunar Orbiter II image from the Moon in the vicinity of the crater Cayley B in the Sea of Tranquility. In it, there were objects casting extremely long shadows that seemed to imply that the objects themselves were "towers" of seventy feet or more. Such objects, if they really were present on the lunar surface, would almost by definition be artificial. Eons of meteoric bombardment would have long since blasted any such naturally occurring objects into dust.

The "spires" were first reported on by Thomas O'Toole in the Washington Post the day after they were photographed. A subsequent article in Newsweek magazine added to the intrigue, and William Blair, a Boeing Company anthropologist, was the first to closely scrutinize them. Blair had extensive experience examining aerial survey maps to look for possible prehistoric archeological sites in the Southwest United States. In articles for the *Los Angeles Times* and then the *Boeing News*, Blair noted that the "spires" had a series of contextual, geometric relationships to each other. "If such a complex of structures were photographed on Earth, the archeologist's first order of business would be to inspect and excavate test trenches and thus validate whether the prospective site has archeological significance," he was quoted in the L.A. Times.

LO frame 67-H-218

Lunar Orbiter image LO2-61H3 (NASA)

The response from Dr. Richard V. Shorthill of the Boeing Scientific Research Laboratory was swift. "There are many of these rocks on the Moon's surface. Pick some at random and you eventually will find a group that seems to conform to some kind of pattern." He went on to claim that the long shadows were caused by the fact that ground was sloping away from relatively short objects, thereby elongating the shadows.

Blair's rebuttal would later put Shorthill's arguments in their appropriate context: "If this same axiom were applied to the origin of such surface features on Earth, more than half of the present known Aztec and Mayan architecture would still be under tree- and bush-studded depressions—the result of natural geophysical processes. The science of archeology would have never been developed, and most of the present knowledge of man's physical evolution would still be a mystery."

Subsequent analysis seemed to indicate Shorthill was wrong on all counts.

In 2001, Lan Fleming of a group calling itself the Lunascan Project, conducted an analysis of the "Blair Cuspids" as they were now known and concluded that the long shadows were caused by

34

tall, spire-like objects. Later analysis indicated that the objects might not be so exceptional after all, and despite Blair's "limited and highly speculative analysis of suspect coordinate relationships" that seemed to indicate the objects were distributed according to tetrahedral geometry, the general consensus today is that the "Blair Cuspids" are fairly normal boulders that were photographed under unusual lighting conditions. Current Lunar Reconnaissance Orbiter images would seem to support that conclusion.

Lunar Orbiter III was the most prolific of the landing site missions, taking nearly 700 medium and high resolution images of the lunar surface from about 30 miles up. The final two missions, *Lunar Orbiters* IV and V, were high altitude missions designed for overall lunar mapping purposes. All of the Orbiters were subsequently crashed into the lunar surface intentionally to study tracking capabilities and measure the impacts themselves.

Having paved the way for landing site selection, the Lunar Orbiters could be set aside for the far more important Surveyor series. Surveyor, it turned out, was critically important because until one actually landed on the Moon's surface, no one was quite sure what they would find there. In the 1950's, Thomas Gold, an Austrian-born astrophysicist and professor of astronomy at Cornell University, speculated that the lunar regolith (the powdery dust that covers the entire lunar surface) might be as much as 10 feet deep in some places, making a safe landing on the surface well-nigh impossible. Gold later revised his assessments, and his revised prediction of the depth of the Moon dust (1-2 inches) turned out to be uncannily accurate. But his studies certainly concerned NASA enough that the Surveyor program was considered a high priority.

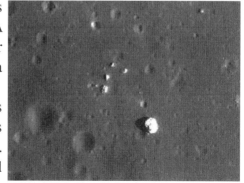

The Surveyor missions also served other purposes that were equally critical. While the Soviet's had achieved the first soft landing

The "Blair Cuspids" today? (NASA/LRO)

35

on the Moon in 1966 with Luna 9, the U.S. had not done so and Surveyor was designed to test the landing radar systems and retro-rocket technology that would be used to land men on the Moon itself. Equipped with a high resolution television camera, Surveyor would be the first U.S. spacecraft to broadcast live images (as still photographs) from the Moon's surface.

Surveyor I was launched on May 30th, 1966 on a direct landing trajectory (meaning it would not orbit the Moon, but would simply proceed on a collision course and then fire breaking rockets to land). It took close to 3 days to reach the vicinity of the Moon, and then fired its retro-rockets, slowing the spacecraft to a descent and soft landing in the Ocean of Storms (Oceanus Procellarum) on June 2nd of the same year.

Although Surveyor's camera was black and white, color images of the lunar surface could be generated by taking three images with three different color filters (red, green and blue) applied over the lens. On Earth, the three images could then be overlaid to produce a single full-color image. When the images were created, they showed – to the scientists' surprise, that the lunar surface is actually a very colorful, if not multi-colored place. Quickly deciding they must have done something wrong, the scientists "corrected" the color to produce flat, gray looking surface features. It was only decades later that this mystery made sense and the weird multi-colored environment of the lunar surface could be explained. But that's a later chapter.

After the success of Surveyor I, Surveyor II quickly followed suit but became the first major setback for the program after one of its breaking rockets failed to fire during a mid-course correction. The spacecraft began tumbling and instead of soft landing in Sinus Medii (the "Sea in the Middle") it impacted at high velocity (about 6,000 miles per hour) near the crater Copernicus and was completely obliterated.

Surveyor III was more successful, landing in Oceanus Procellarum on April 20th, 1967, but was not without its own problems. Two of the landing retro-rockets failed to shut down as planned and the spacecraft "hopped" twice on the surface of the Moon before settling into its final resting place. The Surveyor III spacecraft was later used as a landing target for the Apollo 12 mission in 1969 and the Lunar Module Intrepid successfully landed only 600 feet away. It was also the first mission to have a soil sampling scoop to analyze the lunar surface.

All of this only served to set up Surveyor IV, which should have been the crown jewel of the Surveyor program. After the failure of Surveyor II, Surveyor IV was scheduled to land in Sinus Medii and fulfill the lost mission of its sister ship. Sinus Medii was at the time considered the favorite for the first manned lunar landing, primarily because of its central location, relatively flat surface and interesting topography. All that was needed was for Surveyor IV to land and return some vital "ground truth" of the possible landing area. It didn't happen.

While descending to the lunar surface on its terminal-descent phase, the spacecraft simply *disappeared* some 2 and ½ minutes before touchdown. Since the solid fueled descent retro-rocket was only some 2 seconds from cutting off, NASA publically concluded that the vehicle must have exploded at high altitude. However, if this was the case, a slow chemical explosion would have been recorded on the spacecraft's telemetry sensors, allowing NASA to reconstruct the events as they happened. There was no such telemetry. It simply ceased to exist from one moment to the next.

While this must have disturbed everyone at NASA, eventually they were able to land Surveyor VI in Sinus Medii. But what that spacecraft subsequently found must have gone a long way towards explaining why we never attempted a manned Apollo landing there, and to explaining what really did happen to Surveyor IV…

Television image of shadow of Surveyor 1 footpad as it descends to the lunar surface. (NASA)

CHAPTER THREE

SINUS MEDII

Almost as soon as the U.S. probes began sending back images, there sprang up a small cottage industry of investigators looking at pictures and finding the unusual. Some of these efforts were well intended but amateurish; others were so bad that they seemed almost designed as disinformation. The first book calling attention to possible artifacts on the Moon was George Leonard's *Somebody Else is on The Moon*. Published in 1976, Leonard's book was a collection of his observations of various oddities that he saw (or thought he saw) in numerous Lunar Orbiter and Apollo photographs of the lunar surface. For the most part, his ideas were quaint but interesting, as he pointed to all manner of "cranes," "towers" and "X-drones" that he claimed were actively mining the surface of the Moon. What is most interesting to me however was that Farouk El-Baz, an Egyptian geologist who was the head of NASA's manned landing site selection team, actually met with Leonard to look at what he'd found. I've often thought that he did this to perhaps not only see what he and his team had missed in terms of Ancient Alien evidence, but also to see where they had slipped up in *hiding* that

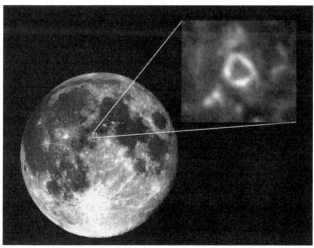

Close-up of Ukert crater (inset) from North American catalog

39

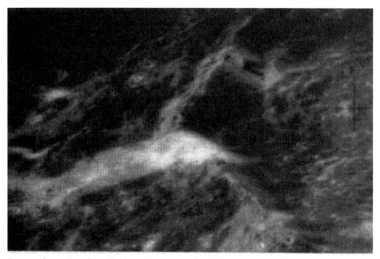

Image from *We found Alien Bases on the Moon* purportedly showing cloud cover on the lunar surface. More likely, Steckling is seeing a glass dome structure acting like an atmosphere over the surface.

evidence.

Another well-known comedy of errors is Fred Steckling's *We found Alien Bases on the Moon*. In it, he points to various developer stains and photographic defects as "proof" that the Moon has an atmosphere, lakes, vegetation and cloud cover. Now, some of the images do show various "mists" and "fogs," which as it turns out, are anything but. However, what they turned out to be is so much more interesting...

Still, it's hard to be too critical of these early efforts by lunar anomaly hunters. The Moon presents quite a challenge for any investigator. The biggest problem with looking for evidence of ancient ruins on the Moon is not so much *what* to look for, but *where* to look. Or more exactly — where to *start* looking. The Moon is a big place, with a land area larger than the African continent, and there are literally tens of thousands of photos from the various NASA programs like Lunar Orbiter, Surveyor, Ranger and the manned Apollo missions. Deciding where to start looking would be a daunting task for anyone.

Luckily, in the early 1990s, researcher Richard C. Hoagland (my co-author on *Dark Mission*) got a big break that made it is easy for him to figure out where to look and what to look for. A friend had given him a 1960s era North American Aviation catalog made up of photos of the Moon taken by Earth based observatories. Having already spent a good number of years working on NASA images of Ancient Alien ruins on Mars, Hoagland had a pretty good idea what he was looking for and quickly spotted his first clue as to where to begin his investigation.

At first glance, each photo resembled the next—distance and close-up shots, craters and maria. But then he looked in the corner of page 241 of the catalog, where a photo of the area around a crater named Trisnecker first appeared in the collection. There, right next to Trisnecker, was one really weird looking crater. This weird crater (named "Ukert," after a German scholar) was not only triangular in shape, but its sides were made up of bright, highly reflective material while the darker center made up the geometric shape. It was almost as if someone had framed the triangle to help us spot it.

But, was it "a trick of light and shadow," as NASA likes to say, or was it real?

This is all even more interesting when one considers that Ukert is found at almost the exact center of the lunar disk as viewed from Earth. At times, this 16 mile-wide crater is directly opposite, or under (if you're standing on the lunar surface) the "sub-Earth point," the location on the Moon where the Earth would be seen directly overhead. So, a near perfect triangle located right in the middle of the Moon? It was almost as if someone

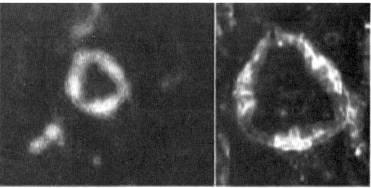

Two views of the triangular shaped crater Ukert from Earth based Lick Observatory (left) and NASA's Clementine probe (right).

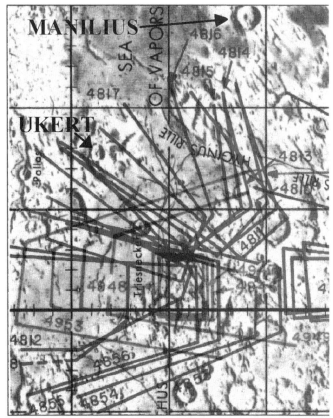

NASA footprint map of Sinus Medii reconnaissance photography.

wanted us to find the area and look around.

Ukert is located in a region known as Sinus Medii, Latin for "The Sea (in the) Middle," and as I mentioned in chapter 2, early in the Apollo program it was considered the first choice for a manned lunar landing site when the Apollo missions began. But as we will see later in the book, NASA's photographic exploration of the area must have quickly scared them off that idea...

By the time the practice landing mission Apollo 10 was launched in May 1969, Sinus Medii had been rejected as the site for the first full-scale manned lunar landing, scheduled for Apollo 11 in July, 1969. But this didn't stop the Apollo 10 astronauts (Tom Stafford, Gene Cernan and John Young) from taking hundreds of pictures of the Sinus Medii region with hand held Hasselblad cameras.

It's obvious from the footprint maps that someone at NASA

wanted to photograph the hell out of the entire Sinus Medii/Ukert region, in spite of the fact NASA no longer had any intention of landing there. The question is why? Many of the photographs, like frame AS10-32-4819, seemed to focus less on the flatlands where a landing might be made, and more on the weird geometric areas and mountains where such a landing would be pretty much impossible. Was NASA looking for something else besides a future Apollo landing site?

Encouraged, Hoagland began by ordering pictures from some of NASA's early 1960's robotic probes sent to the Moon, especially the Lunar Orbiter series of spacecraft. He quickly found evidence of something far stranger than simply rocks or mountains.

The Shard

The first photograph from this probe that Hoagland was able to acquire and examine (LO-III-84M) immediately revealed a number of striking, if bizarre objects on the photos. As he was scanning the horizon of LO III-84M, he noticed something odd. There, sticking straight up more than a mile high out of the lunar surface, was something that absolutely did not belong; The "Shard."

Wide angle view of the Shard, sticking 1.5 miles above the lunar surface. Bright smudge beyond is The Tower, a similar structure more than 7 miles high. "X" shaped object is a frame alignment registration mark. Note the shadow being cast across the lunar surface by the Shard, proving it is not a photographic defect or enhancement artifact.

On a bashed and battered lunar surface, exposed to billions of years of unshielded asteroid and meteorite bombardment, the Shard was a highly unusual, defiantly upright bowling-pin shaped structure, with an irregularly-pointed apex, a swollen middle "node" and a narrowing "foot." Simply put, after billions of years of meteoric bombardment, an object like the Shard simply *can't* exist anywhere on the surface of the Moon. Just like the Blair Cuspids, the very existence of its long shadow could serve as proof of its artificiality. With no atmosphere to protect such a structure from the constant onslaught of meteors, it should have been ground to dust eons before, even if it was some sort of exotic natural structure. Its very existence there on the Lunar Orbiter photo is in of itself proof of its artificial nature.

A closer look at the photo shows that it is not some sort of photographic defect, either. The readily-apparent shadow it was casting across the lunar surface is consistent with the local geography on which the Shard was sitting. In addition, the shadow was also consistent with the time of the month, the Lunar Orbiter camera angle, and the actual sun-angle illuminating the object when the image was initially acquired in mid-February 1967. It was this shadow, more than any other single aspect of this object, which solidified the Shard's reality as an extremely anomalous,

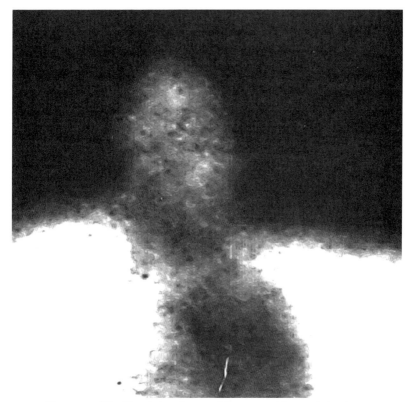

Close-up of the Shard showing semi-transparent internal structure.

potentially manufactured lunar structure.

There were other, crucial clues that the Shard was, in fact, a real object–still standing upright against gravity on the lunar surface. One was its alignment with the local vertical rather than with the grain of the Lunar Orbiter film; the second, was a highly geometric internal organization. But what was it made of? The whole thing seemed to shimmer in the light like some impossible Crystal Tower of the Moon.

The striking geometric pattern was composed of a repeating, complex, internal crystalline geometry, visible all throughout the object. Additional enhancements revealed this regular, internal pattern was made up of highly reflective and possibly hexagonal geometric compartments, greatly damaged but still visible. The overall impression one was left with was that of a once-much-larger, complex, crystalline, artificial object now extremely

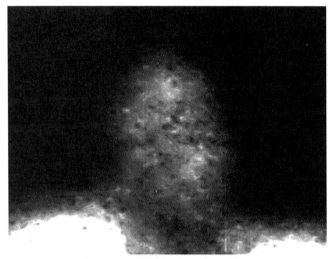

Ultra close-up of the top of the Shard.

eroded by eons of meteorite impact processes.

Since it is virtually impossible to argue with the shadow cast by the Shard, and since the object's very existence (therefore established) argues forcefully for its artificiality, some critics have looked to alternate explanations for its presence on LO-III-84M. It has been suggested, for instance, that the Shard may be a transient out-gassing event, luckily captured by the Lunar Orbiter camera. But, the absence of any diffusion or spray around the Shard's sharply defined edges and all that obvious internal geometry works heavily against this explanation.

But there was an even more compelling argument favoring the Shard's reality, to be made right from the same LO-III-84M frame: the striking presence of "the Tower"…

The Tower

Just to the left of the Shard in the Lunar Orbiter frame LO-III-84M lies what at first appeared to be a faint smudge on the image. However, on close inspection and under additional enhancement, it quickly became obvious that this is a second glass-like anomaly in the same frame. Christened "The Tower" by Hoagland because of its immense size compared to the Shard, the Tower appears to be rising from a point over the horizon, making it at least 260 miles from the camera and more than 7 miles high! Like the Shard, it is aligned with the local vertical rather than the camera or the film grain, and there are even hints of guy-wire like filaments holding the immense structure up.

But could there be another explanation? Maybe it was a comet, or a star or even a far distant nebula or galaxy? The comet idea was quickly discarded when a quick check found no comets of any significance passing through the solar system in February, 1967. As to the possibility of it being a star, nebula or distant galaxy, those possibilities were also put to rest rather easily.

NASA's published data on Orbiter photography showed that the f-stop setting for this image was 5.6, the shutter speed 1/100th of a second. Simultaneously capturing the bright lunar landscape and a faint background interstellar nebula on the same photograph, with those specific camera settings, on this type of slow-speed, fine-grained film, would be technically impossible. If the camera aperture was held open long enough to capture such an incredibly faint object, the brilliant, sunlit lunar landscape would have been completely blown out. With these possibilities eliminated, the other mundane explanation, that the feature was merely a photographic blemish, was also eliminated through independent analysis by photographic experts at a photo lab in New York.

The bottom line, whatever the Tower was, it had to be local, to the Moon at least.

Analysis of the Tower itself showed it was more than a mile across and was an estimated distance of at least 300 miles from Lunar Orbiter when the image was taken. Just like the Shard, the top of the Tower itself appeared to be composed of dozens of smaller cubical (and/or hexagonal) sub-structures. And these were clearly not artifacts of the enhancement process either. The smallest

The Tower (left) and the Shard (right).

47

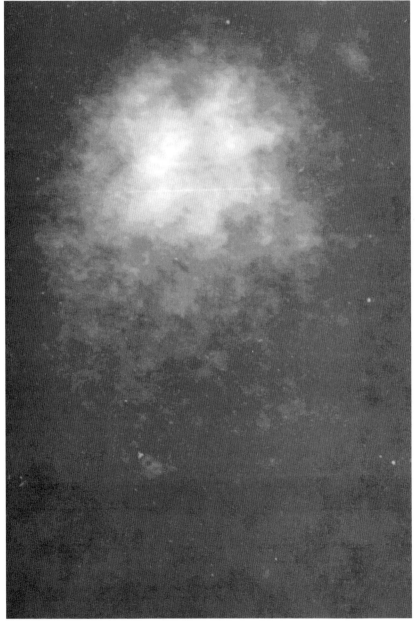

The Tower (close-up).

of these cubes visible on frame III-84M measured at least 50 times the size of the individual computer pixels of the imaging enhancements.

In other sections of the image, long, vertical transitions created

48

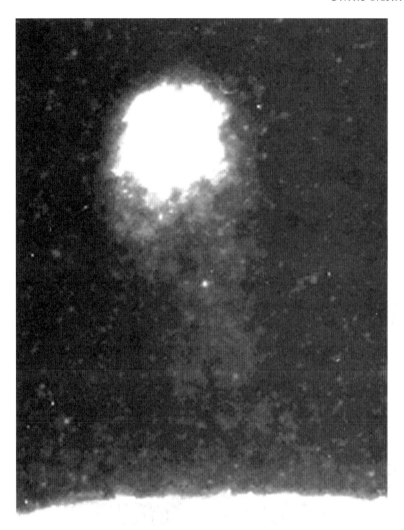

The Tower (enhanced).

by apparent refraction-effects could also be traced, as if portions of the background object were being viewed through a heavily-distorting medium, located much closer to the spacecraft. The Tower also appeared to taper towards that surface, and simultaneously, to be leaning in a southerly direction in the photograph. This obvious departure from the local vertical (similar to the internal details in the Shard) could be directly connected with the distance of this object from the Lunar Orbiter cruising thirty miles above the Moon, looking downward toward the lunar horizon. A real object,

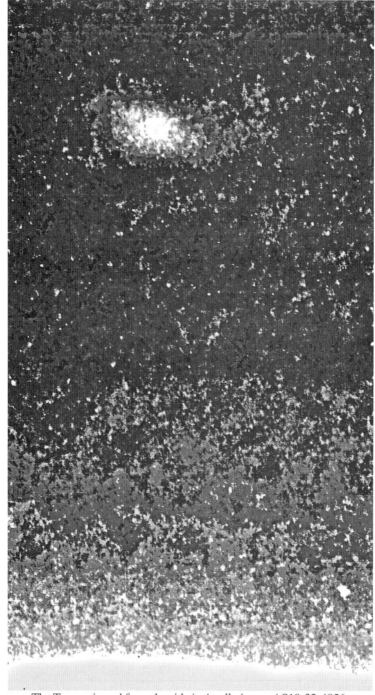

The Tower viewed from the side in Apollo image AS10-32-4856.

connected with the surface by an actual vertical tower, could indeed appear to lean, if (a) it was located closer to the spacecraft than the Shard, or (b) was in the process of slowly tipping over from the effects of constant meteor erosion, across literally millions of years.

Not only is all this patently absurd within any current (or proposed) natural geologic model of the lunar surface, it totally supports the concept that the Tower/Shard are nothing but remaining fragments of a once far-larger, also clearly artificial structure. This structure, because of its transparent nature, was apparently once composed of something like glass, and, attached to some kind of darker, vertical structural framework.

After confirming the existence of the Tower from another photograph taken by the Apollo 10 mission, it became obvious that there was some kind of transparent, glass-like structure over large swaths of the Sinus Medii region. But how could such a structure survive on the lunar surface for so long? Isn't it dangerous to build glass houses in places where people throw stones (like the Moon)?

That question is actually easy to answer: all materials, including glass-based minerals like quartz silica, take on different properties in the hard, cold vacuum of space than they do on Earth. One of the most abundant substances of this world is water; in liquid, gaseous or solid form. Space however, has an appalling lack of water, so much so that it can be easily stated that water is one of the least available resources in the vast expanse of emptiness we call space.

As it turns out, this is one of the qualities of the vacuum that makes glass not just a desirable, but an *ideal* material for building structures on an airless world like our Moon. Glass on Earth is well known to have little tensile strength, meaning it doesn't stretch easily because it is brittle and will not withstand even a very weak impact from a hard object. When you throw a baseball at a glass window, it fractures and cracks easily and with little resistance. However, if you attempt to crush a glass sphere, you'll find that it has a great deal of strength under compression stresses.

The reason for these properties on Earth is that it is pretty much impossible to extract the water from glass as it is forming. Water is all around us, even in the most arid deserts. It is in the ground as a liquid, frozen in the arctic as a solid, and even in the air around us as humidity. All this water causes a phenomenon called "hydrolytic weakening" when glass is being manufactured on Earth, meaning that at the molecular level, the bonding of silicates and oxygen is resultantly weakened. This produces a transparent, brittle and fragile material we call common glass. Manufactured under Earthly conditions, we have found that glass is a very useful and artistically pleasing medium for a variety of uses -- but structural construction is not one of them.

In short, on Earth, we don't build glass houses.

But the Moon is a completely different story. It is airless, with no humidity to interfere with the molecular bonding of the silicates that make-up the glass that is omnipresent. The hard-cold vacuum enhances the strength of lunar glass to the point that it is approximately twice as strong as steel under the same stress conditions. In fact, several papers from scientists at Harvard and other universities have suggested that lunar glass is the ideal substance from which to construct a domed lunar base.[1]

All I'm proposing is that somebody else came up with the idea long before we did. "Somebody," as in Ancient Aliens.

But if this concept is valid, then there must be other evidence to support it, right? As it turns out, there's plenty.

Surveyor 6

Surveyor 6 was an unmanned NASA robot probe that successfully landed on the Moon in November 1967 about thirty miles west of "Bruce," a small five-mile wide crater near the center of Sinus Medii. From there, the Surveyor spacecraft took over 35,000 images of the surrounding lunar landscape. One night after local sunset on November 24, an additional set of time-exposed images were acquired looking west, for purposes of studying light-scattering properties in interplanetary space caused by the solar corona, which was at the time far below the lunar horizon.

One of the Surveyor 6 images.

These images, when they were released, caused something of a fire storm.

Instead of the Sun's faint corona, the image contained a huge flare of light on the horizon, very similar to a common sunset here on the Earth. The problem with this lunar sunset was that it was anything but common. The Surveyor image, with its remarkably brilliant beads of light stretching along the western horizon and an intensely geometric structure of scattered light seen against the lunar sky above it, was taken over an hour after sunset.

On Earth, when we see the sun set, the sun in fact has already "set" (dropped below the visible horizon) several minutes before. But the atmosphere of the Earth bends the light coming from the sun, so the sun we are watching slowly disappear over the horizon is actually already several degrees below that horizon. This atmospheric interference is also the reason the sun tends to turn more orange and distort in shape as it sets.

All of this is well and good, except for one problem; the Moon has no atmosphere. So

Surveyor 6 image.

53

when *Surveyor* took the image, all it should have seen was the faint wisps of the Sun's descending corona. This was certainly 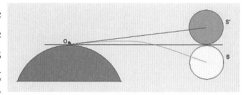 what the scientists expected to see.

The NASA guys quickly tried to come up with an explanation, and they settled on the idea that electro-statically charged fields had somehow suspended particles of dust in the lunar sky and created the dramatic sunset effect. This theory was later discounted

Scaffolding on the Moon. Close-up of box like, semi-transparent structures miles above Sinus Medii. Note the 3D structure and depth, and the interlocking sections of supports and structure. (NASA frame AS10-32-4816).

by lunar soil samples, but the NASA explanation has never been officially withdrawn.

Which left a major mystery; what could cause such intense illumination more than an hour after sunset? The only viable answer is that there must have been some sort of transparent, intervening medium at (and beyond) the horizon that bent the Sun's light and created the dramatic sunset effect. That "intervening medium" is nothing less than our theorized glass dome over Sinus Medii.

In fact, close-ups of the sunset images revealed this much fainter, lattice-like structure arching far above those mysterious brilliant beads along the horizon. This fading light was scattered just the way an atmosphere would scatter it. Only this "atmosphere" had geometric structure.

These optical phenomena could only be caused by the remnants of some kind of ancient, nearly transparent, glass-like, highly geometric architecture still anchored in the lunar surface and extending literally miles above it. The interlocking, scaffold-like structure revealed in the Surveyor 6 images is very similar to the miles-high Shard and Tower elsewhere in Sinus Medii, but much better preserved. What was all but beaten to a pulp in the area of the Shard and Tower is remarkably intact just over the horizon some 100 miles distant.

Later, a further confirmation of this Sinus Medii "dome" was found on a picture taken by the Apollo 10 crew, NASA frame AS10-32-4816.

Fortunately, the location of the Sinus Medii "dome" was amenable to further research, as in searching the image archives it turned out that a great deal of it had been photographed on later missions, especially Apollo 10.

AS10-32-4822 – Los Angeles

After it was clear that Apollo 10 had taken a ton of photographs of the area where Hoagland suspected an immense scaffold-like structure had been constructed, he and other researchers began to pour over the Apollo 10 photo catalogs in an effort to find further images of the "dome," Ukert, and Sinus Medii in general.

AS10-32-4821   AS10-32-4822   AS10-32-4823

As these photographic catalogs were analyzed, a strange pattern quickly emerged. A great many of the thumbnail images in the catalogs appear to be very dark or almost completely blacked out, as if the astronauts had taken most of the photos with the lens cap on. This would tend to discourage an investigator from ordering prints of these images, since getting them from the NASA archives is not exactly cheap. So of course, being the trusting type, what did they do?

They ordered the blacked out pictures of course.

When you actually see some of the prints, it becomes obvious not only that the pictures are perfectly good, but that NASA would have very good reasons to discourage anyone from taking a close look at them. Some of these Apollo 10 images are absolutely chock-full of the kinds of objects that confirm the Sinus Medii dome model. Foremost among these was Apollo 10 frame AS10-32-4822.

The first thing that stands out about this image is a geometric crisscrossing pattern on the surface just northeast of Ukert that

AS10-32-4822 (NASA/Bara).

AS10-32-4822 close-up (NASA/Bara).

seems to defy conventional explanation. One consulting geologist (Dr. Bruce Cornet) even nicknamed the region "L.A. on the Moon" because of its strikingly urban appearance.

In the photograph, over an area of hundreds of square miles (roughly equivalent to the real Los Angeles basin on Earth), there appears a remarkably regular, rectangular, raised, repeating 3D geometric pattern. Large surface grooves stretching for tens of miles appear remarkably similar to streets running across the actual L.A. basin in Southern California. Here and there, small round craters cut into the areas of sharply contrasting, remarkably rectangular relief, like mile-sized cookie cutters. In a close-up from 4822, this rectilinear, artificial block-like pattern, interspersed with a smattering of remarkably uniform impact craters, is even more apparent. Overall, the overwhelming impression is that of finding a vast, ancient, bombed-out city on the Moon.

Even more bizarrely, after ordering the same picture (AS10-32-4822) from seven different NASA archives, it was eventually determined that there were no less than *nine different* versions of the same NASA image, all stacked under the same photo-frame number in the catalog. What this meant is depending on where you ordered your blacked out photo from, you would get a completely different picture, with different lighting and visible detail. What we eventually determined was that these different 4822's were all part of the same sequence of images taken with a power winder camera as Apollo 10 passed over this region of Sinus Medii. Why this was done became obvious fairly quickly. Putting the vari-

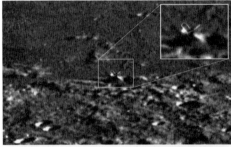

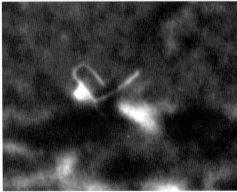

The "Paperclip" (close-up).

ous 4822's side-by-side, some objects (because of the change in the lighting angle) become visible and then disappear. And none of them look like natural objects at all.

Within this general, very artificial-looking landscape, a series of smaller, very brilliant, horizontal, vertical and near-vertical features also appear. Some are clearly resolved as rectangular structures, others as possible sky-scrapers. Other features are seen as merely brilliant, geometrically arrayed points of light—possibly specular reflections from surviving optically flat areas similar to windows or entire glass walls.

One of my favorites of these is an object at the edge of the Los Angeles basin I call the "Paperclip." It has a central post more than a mile high (and which is casting a shadow on the lunar surface) and some obviously metallic antennae which are also casting a shadow in the surface. Whatever this thing is, it isn't of natural origin.

Later, just before the publication of *Dark Mission*, another researcher named Steve Troy found a new series of images of the Los Angeles area, and what these images showed was even more compelling. One such image (AS10-31-4652) revealed layer after layer of obviously geometric glass, reflecting the brilliant light of the rising sun almost directly back toward the approaching spacecraft. One can clearly see multiple layers, floors, and innumerable right angle views of what can only be geometrically-arranged manufactured structures. There are even what appear

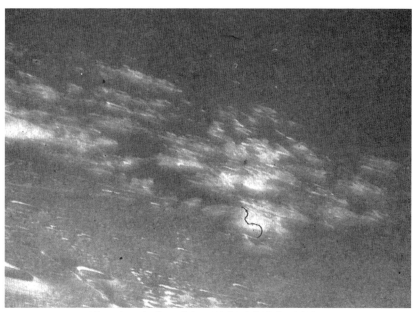

"Los Angeles" at night (NASA AS10-31-4652).

to be suspended walkways and transportation bridges clearly visible.

There is simply no plausible explanation for these incredibly reflective, geometrically aligned, transparent structures except an artificial one. But beyond "L.A.," which was bizarre enough, the real prize of 4822 turned out to be what, at first, appeared to be just a small scratch on the negative.

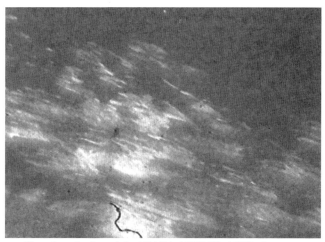

"Los Angeles" at night (NASA AS10-31-4652) close-up.

"The Castle"

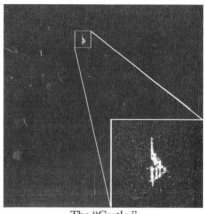

The Castle is another glittering, remarkably intact geometric formation first identified on a leaked version of frame 4822 provided by a highly-placed source at NASA's Goddard Spaceflight Center in 1992. Bearing a striking resemblance to "Schloss Neuschwanstein," a castle built by King Ludwig II of Bavaria in 1869, (which served

The "Castle."

as the model for Cinderella's Castle in Disneyland) it is, in fact, another Ancient Alien artifact hanging high above the Moon.

The Castle's location on the lunar landscape is as remarkable as its appearance. As judged by the geometry on frame 4822, this highly anomalous object is actually suspended some *nine miles* above the lunar surface, somewhere between the eighteen-mile diameter crater, Triesnecker and the well-known Hyginus Rille.

Eventually identified on two separate NSSDC versions of 4822, the Castle raised a series of extraordinary puzzles, beginning with the obvious; what is holding it up?

Stereo analysis of these two versions confirmed the Castle's presence miles above the lunar surface, apparently just hanging out in space. Around it, as previously noted, appeared a collection of much smaller, equally reflective geometric slivers, as if they all were simply fragments of a once much larger, previously intact, somehow suspended structure.

Like the Tower, the Castle too is surrounded by a faint matrix of sparkling geometric, aligned structure. There is also the clear implication of a sagging support cable seen at the very top of this amazing artifact, to which the obviously large and massive structure is physically attached. In other words, the Apollo 10 crew, via 4822, apparently recorded another section (far northeast of the Tower/Shard remains) of this extensive Sinus Medii dome.

Another interesting point about the Castle is that in the original

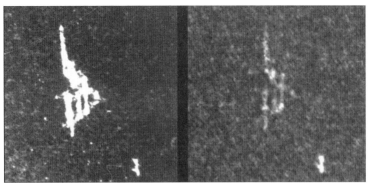

Two distinct views of the "Castle" from two different versions of NASA frame AS10-32-4822. Note the wire passing through the tip of the Castle and sagging under its weight.

version (above—right), the drooping cable is clearly visible; but in the second version of 4822 (above—left), not only has the cable disappeared with the increased viewing angle, but the entire structure has visibly foreshortened. This is the result of optical parallax, as the Apollo spacecraft moved farther to the west in the few seconds between the two exposures. From this fact alone, there is no doubt that, despite sharing the same frame number, these are

two *separate* views of the same object, present on two distinctly different photographic images. And that fact alone makes it a sure fire bet that it is really there, hanging miles above the surface of Sinus Medii.

One of the biggest problems with this evidence of an Ancient Alien presence on the Moon is that the theorized lunar "dome" is just so hard

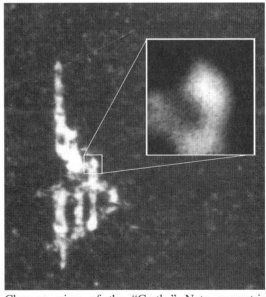

Close-up view of the "Castle." Note geometric internal structure and drooping support cable holding the structure up.

61

for people to visualize. When I say dome, I'm not talking about a dome in the traditional sense like the domed stadiums we are so familiar with on Earth. What these structures really look like are tall, layered box-like structures designed to protect

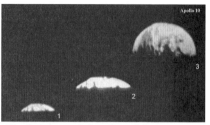

Earthrise over Mare Smythii sequence.

the vast cities below (like Los Angeles) from meteor bombardment. Again, this isn't easy to picture, but back in the 1990's award winning architect Robert Fiertek attempted to do just that.

Using all the available images that seemed to show evidence of this glass structure (including all the multiple versions of 4822), Fiertek painstakingly mapped and entered all of the visible points of light into a ray tracing program. He was inspired to do this because of a project he had conducted with 16mm footage of an Earthrise over a region of the Moon named Mare Smythii.

In this footage taken from the Apollo 10 mission, the Earth can be seen rising from behind the lunar horizon. But for some reason, the Earth looks distorted, flattened and compressed. This might not seem so unusual, since the Moon looks the same way when viewed from Earth.

Moonrise in the Earth's atmosphere.

But again, the problem here is that the reason the Moon looks flattened and distorted is that the Earth has an atmosphere that bends the light and compresses the image of the Moon from the viewer's perspective. But since the Moon has no atmosphere – at least none to speak of – this cannot be the reason that Earth looks so distorted as it rises. Once again, there has to some nearly transparent medium between the camera and the Earth to create

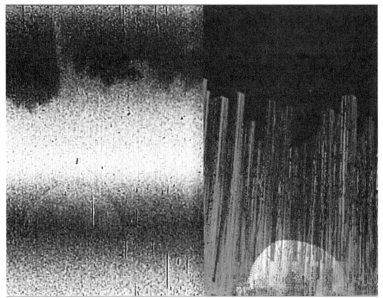

Final, overexposed frame from Earthrise over Mare Smythii film (left) and from Fiekrtek's computer simulation of the transparent lunar dome theory (right).

this effect.

Fiertek quickly realized that he had chance to disprove the entire lunar dome theory, so he set out to reproduce the effect with computer modeling. Plugging in his best guess for what this dome might look like, he was able to reproduce the distortion most effectively.

Theorized structure of the glass dome over Sinus Medii.

Taking this 3D model and tying it back in with the light source data he'd already entered from the Sinus Medii photos, Fiertek was then able to generate a viable computer simulation of the structure of the dome itself. The results were astonishing.

So this dome, as it were, is more like layer upon layer of steel-hard, transparent material, probably glass, protecting vast stretches of city-like infrastructure below. It must have taken eons, truly millions if not billions of years, for this immense and incredible feat of engineering to have eroded to the point it had by the time we went to the Moon and rediscovered it. As you will see, this unbelievable accomplishment is just the tip of the iceberg. There is much more to be found on the Moon, left behind and placed there by who knows who.

I'm not saying it was aliens... But...

What happened to Surveyor 4?

What all of this leads us back to is Surveyor 4. If you remember, that was the robotic mission that disappeared while descending at high speed over Sinus Medii. The Surveyor 6 spacecraft that managed to land in the same region of the Moon safely and take the astounding after sunset lunar dome image from the surface was a replacement for this lost mission. In fact, it wasn't until after that iconic photograph that NASA suddenly decided that Sinus Medii was suddenly unsuitable for the first lunar landing, even though it had been the first choice all along. What I strongly suspect is that NASA actually figured out what must have happened to Surveyor 4.

In short, it went "splat" like a bug on a windshield.

The evidence to support this contention is strong. First, keep in mind that the spacecraft was descending at several thousands of miles per hour at least, given that it was still over two minutes from its designated touchdown site. At that speed, even a minor scrape with an object like the Tower or the Shard would have destroyed the space vehicle. Second, Surveyor 6, rather than land in the same place and use the same approach vector, took a different approach vector and landed about ½ mile from where Surveyor 4 was intended to touch down. Another tell-tale sign is the mysterious way in which the spacecraft simply ceased to exist. One microsecond it was there, the next it wasn't. As I pointed out before, if there had been a chemical explosion, as NASA suggests, then the telemetry would have recorded this rather slow moving process as one system after another on the spacecraft was destroyed by the chemical explosion. Instead, the spacecraft simply blinked out of existence. Also, if NASA truly thought there was a malfunction with Surveyor's descent rockets, wouldn't they have tested that thesis and made some improvements to the system? In reality, they did not.

This disappearance then can only be explained by one of two possibilities. Either Surveyor 4 was sucked into some sort of hyper-dimensional wormhole created by the Moon's Ancient Alien defense systems, or it hit something so fast that not even the light-speed radio transmissions from the onboard computer had time to record it. For obvious reasons, I favor the latter scenario.

At first, it may seem implausible. After all, wouldn't the spacecraft's sensitive telemetry also record the considerably slower-than-light impact against the towering glass structures we've speculated are littering the Sinus Medii region? Theoretically, yes they would. But the cool thing about conventional physics is there are always exceptions to virtually every rule. A case in point; the "scissors effect."

Put simply, the scissors effect is an example of the fallacious idea that nothing can travel faster than light (see *The Choice*). It says that if a pair of scissors were suddenly closed at speeds approaching light speed (1.0c), then the theoretical intersection point between the two blades would (and must) exceed the speed of light. So if the 2 end points of the blades were closed at 0.9c, the intersection point (which exists only mathematically) would exceed 1.8c, or 1.8

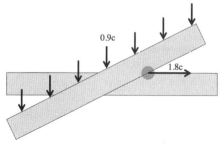

Illustration of the "scissors effect."

times the speed of light. Of course, this calculation ignores the fact that if the two blades actually touch at the point of intersection, as they do in real scissors, then theoretically the blade tips can't get anywhere near the speed of light. But hey, that's a real-world application, and these are physicists so...

Anyway, the point is, stuff *can* exceed the speed of light under certain circumstances. So if, as an example, Surveyor 4 were descending at thousands of miles per hour and it's antenna were suddenly deflected by impacting a giant, immovable object (like say a two times stronger than steel lunar tower extending miles high into the sky above the lunar surface), then the signal might very well be lost so quickly that data about the impact might not be recorded or transmitted back to Earth at all.

Hence, splat! As in a faster-than-light splat.

If any further evidence was needed to prove out the thesis that there is an immense, miles-high, lattice-work type transparent structure above large portions of the Moon, it was provided a few years ago. Shortly before going to publication on our book *Dark Mission*, my co-author Richard C. Hoagland found a new image of the Moon from the Apollo 15 mission that seemed to seal the deal. Apollo 15 photograph AS15-88-12013 was taken by the astronauts from the Apollo 15 command module Endeavour shortly after "Trans Earth Injection," the firing of Endeavour's main engine to break lunar orbit and begin the return trip to Earth. Looking back at the Moon and with the sun behind them, the astronauts snapped a photo with one of the hand-held 70mm Hasselblad cameras they

Apollo 15 photograph AS15-88-12013.

took on the mission. What the photo shows is truly astonishing.

Enhanced Apollo 15 photograph AS15-88-12013.

Under enhancement, the photo shows what appears to be a misty, billowy "air-glow" limb around the Moon, stretching literally miles-high above the surface. Scattering the light into the blue end of the spectrum – exactly as an atmosphere would—it looks for all the world like the Moon has a dense atmosphere. But a closer examination reveals that is the least likely explanation.

All planets with dense atmospheres and solid surfaces, like the Earth, have a visible "air-glow limb." From orbit, it appears as a fairly solid line above the visible horizon, the solid rocky part, of the planet below. The Earth's is strong and distinct, as the above image from the International Space Station shows. But while the Moon's "air-glow limb" from AS15-88-12013 is similar, there are dramatic differences.

For one, it is denser in some places than in others. Notice how the light scattering is thicker above the whiter highland areas in the picture above, and thinner in the areas above the darker maria, or lowland areas. If we were really looking at an atmosphere, it couldn't pick and choose the areas where it decided to be thicker. As the view of the Earth's air-glow barrier shows us, it is visually uniform in density.

So what we're seeing in AS15-88-12013 has to be something else.

It has to be, in short, some sort of transparent (like glass) intervening medium that is scattering the light above and around the lunar disk in this photograph. It has to be our miles-high glass dome, caught under the perfect lighting conditions to capture its gauzy, translucent effect. It has to be the same structures we saw in the famous *Surveyor* 6 photograph, only this time, photographed from lunar orbit (or at least near lunar space).

While all this makes a circumstantial case to reinforce the "miles-high glass dome model" of Ancient alien ruins on the Moon, it doesn't completely seal it. For that, we need the crucial ground truth. Images and or testimony from the astronauts that were actually there, in situ, as they say, on the surface of the Moon and saw these majestic ruins for themselves. Fortunately, we have it. And in droves.

[1] "In Situ Rock Melting Applied to Lunar Base Construction and for Exploration Drilling and Coring on the Moon," Rowley, J. C. & Neudecker, J. W., *Lunar Bases and Space Activities of the 21st Century*. Houston, TX, Lunar and Planetary Institute, edited by W. W. Mendell, 1985, p.465, 1985lbsa.conf. 465R

CHAPTER FOUR
TO THE MOON, ALICE!

At this point in the mid-1990s, there had been a significant movement forward in the study of lunar anomalies. Early researchers like Steckling and Leonard had opened the doorway to close examination of the NASA photographic records of the Moon, and Hoagland had blown it wide open with his findings. But what was missing was that critical ground truth we talked about—direct evidence from the lunar surface that these miles high glass dome structures existed.

Most of the time, when someone uses the word "dome," it implies a watch crystal like, single piece structure over a crater or some other low lying surface feature. Inside such a structure, it would be possible to create an Earth-like environment. But Hoagland's model, giant multi-layered scaffolding type structures, actually makes more sense. A naked, watch crystal type dome would be vulnerable to the kind of high-velocity impacts that Moon has experienced for most of its 4.5 billion year life. But a multi-layered, reinforced glass dome would act almost like an atmosphere in terms of the protection it would provide over eons of smaller impacts and the inevitable degradation they would cause. I'm not saying there aren't "watch crystal" type domes over some lunar craters—in fact we'll study a few later—but they would be the last line of defense for an Ancient Alien civilization on the Moon. The scaffolding would be the first and presumably the most robust. At this point, all that was left was to find some evidence of it—from the ground.

Skyscrapers on the Moon over Sinus Medii and "Los Angeles."

In early 1995, Hoagland was on a lecture tour in Seattle and met Ken Johnston, a Boeing engineer at the time, and a former test pilot for Grumman Aerospace. After his tour of duty in the Marines, Johnston had gone to work at NASA in the mid-1960s as a Lunar Module test pilot at the Manned Spacecraft Center in Houston. There he and his team subsequently trained all of the Apollo astronauts to fly the Lunar Module, while simultaneously being part of the extensive spacesuit development program ("I was 'capsule size,'" Johnston would later joke).

Johnston later moved across the center, going to work for Brown-Root Corporation and the Northrop Corporation in MSC's Lunar Receiving Laboratory (LRL) during Apollo. This consortium had the prime contract for the processing of the actual lunar samples coming back

AS10-32-4862

from the Moon, and Ken's key function was as supervisor of the data and photo control department. This was the section of the LRL that handled all of the critical photographic and written documentation for the Apollo program. After processing elsewhere in the Lab, the films and samples went through Johnston's office for cataloging and long-term storage.

Having read Hoagland's first book, *The Monuments of Mars*, Johnston wrote a letter of introduction and offered at Hoagland a chance to review Ken's collection of about 1,000 old NASA photos and other memorabilia. But the story of just how Johnston came to possess the photographs is very interesting and worth retelling.

As head of the LRL photo lab, it was Ken's responsibility to catalog and archive all of the Apollo photographs taken by the astronauts. As part of the archiving process, the LRL eventually developed four complete sets of Apollo orbital and handheld photography, comprising literally tens of thousands of first-generation photographic negatives and prints. Ken also had responsibility for managing the 16mm mission films from the on-board "sequence cameras" (modified military gun cameras), operating from the Command Module and Lunar Modules during various phases of the missions, including lunar orbit and descent/ascent. One of his duties was to frequently screen these on-orbit films at MSC before members of the various scientific and

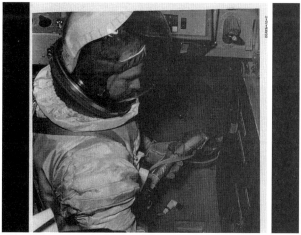

Ken Johnston at the Manned Spacecraft Center in Houston.

engineering teams.

During his time at NASA, he'd had a couple of strange experiences which had nagged at him over the years. One was during just such a screening of the 16mm films which he recounted on the popular *Coast to Coast AM* national radio program:

"Well, on that particular case—this was Apollo 14—after we had received the film, right after the astronauts had returned to the Earth, it had been processed in the NASA photo lab. It was my responsibility to put together a private viewing for the chief astronomer—that was Dr. Thornton Page and his associates and contributing scientists. I took the film over and set it up into what is called a 'sequence [projector]'; it's kind of like one of the gun cameras they use in the military [but in reverse—a projector]—where you can stop, freeze frame, go forward, back up and zoom in.

"And we were viewing the Apollo 14 footage, coming around the backside of the Moon as we were approaching a large crater. Now, due to the sun angle on the front side [of the Moon] that you would be looking at (you'd probably be looking at more of a crescent at that point on the backside) in the shadows in the craters, covering about half the crater, this particularly large crater showed a cluster of about *five or six lights down inside the rim.*

"And this column or plume—or out-gassing or something, coming up above the rim of the crater, where we could see that— at that point Dr. Page had me stop and freeze, and back up; and go back and forth several times. And each time, he'd pause a second and look... and he finally turned to his associates and said: 'Well, isn't *that* interesting!' And they all chuckled and laughed, and Dr. Page said: 'Continue.'

"Well, I finished up that viewing and I was told to check it [the on-board sequence camera film] back into NASA bonded storage in the photo lab. The next day, I was to check it back out and show it to the rank-and-file engineers and scientists at the [Manned Spacecraft] Center.

"While we were viewing it the second time—and, several of my friends were sitting next to me—I was telling them: 'You can't believe what we saw on the backside of the Moon! Wait until you see this view.'

"And, as we were approaching the same crater... and we went *past* the crater—*there was nothing there!*

"I stopped the camera, took the film out to examine it—to see if anything had been cut out—and there was no evidence of anything being cut out. I told the audience that we were having 'technical difficulties,' put it back in and finished.

"That afternoon, I ran into Dr. Page over at the Lunar Receiving Laboratory and asked him what had happened to 'the lights and the out-gassing or steam we saw,' and he kind of grinned and gave me a little twinkle and a chuckle and said: 'There were no lights. There is nothing there.'

"And he walked away. And, we were so busy... I didn't get a chance to question him again."

Johnston had also observed various oddities with the still camera images. Once, while passing through a classified building on the Center he normally didn't frequent, Ken observed artists airbrushing the "sky" in various photos. That in itself wasn't unusual, as press release prints were regularly cleaned up. What bothered Johnston in this case was that these weren't *prints* that were being airbrushed, but rather *photographic negatives*—meaning that, after that drastic process had been applied, the original data could never be reproduced in the form it had originally been taken.

That was bad enough. But what NASA attempted to force Johnston to do later was even worse. As we recounted it in *Dark Mission*:

> ...in 1972, near the end of the manned lunar program, Johnston was called into the office of Bud Laskawa, Johnston's lead at the LRL records division. At the meeting, Laskawa told Johnston that orders had come down from NASA Headquarters (through Dr. Michael Duke, Laskawa and Johnston's NASA boss) to destroy *all* of the copies of the original lunar photography that he had been protecting and archiving for the past several years.

Johnston was dumbfounded that anyone could order the destruction of the official photographic record of Mankind's first venture beyond the earth. He protested, and begged to be allowed to donate the photographs to various universities or foundations, but was told there was "no chance." The orders were explicit—he was to destroy all four sets of the literally tens of thousands of Apollo lunar photos taken by the astronauts.

Johnston found this situation unconscionable. Eventually, after further protests, he relented and destroyed three full sets of the data—but with his guilt eating away at him, he decided to save one complete set "elsewhere." Some of the images and negatives he kept for himself. However, since the collection was so vast, he eventually decided to donate the rest to his alma mater, Oklahoma City University, where the data quietly resided—out of NASA's oversight—for over thirty years...

Unfortunately, when Johnston and Hoagland went to Oklahoma City University and attempted to retrieve the photos, they found that a retired professor had apparently absconded with most of them, leaving only about a thousand first generation prints in Ken's personal collection to examine.

A close examination of Ken's surviving photos revealed overwhelming evidence that the lunar scaffolding model was correct. And it also raised the question of just why NASA would want these nearly priceless photos destroyed.

Because Ken's photos are prints, rather than negatives, theoretically they are of lesser value than what is currently in the NASA archives. But in reality, this is not the case. The original negatives are kept in a sealed vault in NASA's Houston facility, and have only rarely been seen by outsiders. The negatives currently at sites like the National Space Science Data Center in Maryland and in other official NASA archives are in fact multi-generational copies of (theoretically) those originals. In other words, a copy of

NASA photo AS14-66-9301.

copy of a copy, at the very least. Ken's prints however, were first generation, made from those rare original negatives. By that alone, they must have contained more information than the best of NASA's archive negatives do today. And what they showed was astounding.

In going over Ken's collection, a couple of images immediately stood out. The first was Apollo 14 photograph AS14-66-9301, now known as the (in) famous "Mitchell Under Glass" frame.

At first, AS14-66-9301 seems to be a very innocuous photo of the lunar surface. Taken as part of a landing site panorama by astronaut Alan Sheppard, it shows Lunar Module pilot Edgar Mitchell setting up a scientific experiment in the foreground and a substantial amount of lunar "sky" in the background. The interest began when Ken's wife Fran noticed something odd in that background. "Why is the sky *blue*?" she asked.

Looking closer, it was easy to see an odd blue spec on the image in the sky above the landing site. Because of the very short exposure times of the surface photography (on the order of 1/250th of a second) there was no way it could be something in space or

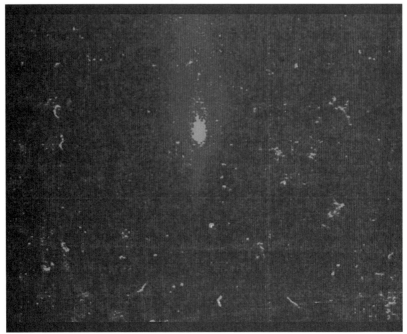

Ultra close-up of specular reflection and geometric structures from AS14-66-9301.

far beyond the landing site. It is a common misconception that you should be able to see stars and other faint, far-off objects in photos from the lunar surface. In reality, the sky should be absolute, seamless black. Unless it was a photographic defect, the blue spec had to be something very close to the landing site itself.

Under enhancement, the blue spec turned out to be a blue-scattered specular reflection off of the same type of towering, lattice-work structure that was seen at Sinus Medii. Since the Apollo 14 landing site was well away from Sinus Medii in the Ocean of Storms (Oceanus Procellarum) it could not be the same structure. It could only be a completely different set of similar but separate transparent, glass-like Ancient Alien ruins.

Further examination of the image and the rest of the panorama revealed that the Apollo 14 Lunar Module Antares had landed literally right in the middle of a vast complex of these towering glass structures in the Ocean of Storms. Not only that, but these structures behaved exactly as they should under light scattering from the sun, brighter near the light source and 180 degrees from it, and darker the farther away from the light source. In other words, if the structures seen in the images were photographic defects, then they were photographic defects that scattered light exactly as if they were real, glass-like transparent structures on the surface of the Moon. But the real "smoking gun" came when the photos were compared to another Apollo mission; Apollo 12.

Apollo 12 had landed on the in the Ocean of Storms some 22 months earlier, at a site only 122 miles away. In theory, a comparison of images from the two landing sites might shed some light on whether these enormous structures were really there on the surface of the Moon. If the same towering, megalithic

AS14-66-9301 showing angled support structure from Apollo 14 landing site.

edifices could be seen images from both missions, then that would be a final confirmation that these were real structures and not any kind of photo defect.

The focus was immediately on the brilliant blue specular reflection seen in the Mitchell Under Glass photo. The source of the reflection seemed to be embedded in a lattice work of filaments and supports at 90 degree angles to the local lunar surface. But behind the reflection was a series of slanted, inclined support structures visibly connecting with lunar surface at an angle. If these massive

AS14-66-9301

and distinctive features could be spotted in some imagery from Apollo 12 as well, it would be the proverbial smoking gun.

An initial search of the Apollo 12 hand held photography was disappointing. The sky in many of the images showed signs of being white-washed (or "black-washed," in this case) and while a few spectacular photos were found, none of them were pointed in the direction of the Apollo 14 landing site, where (theoretically) these massive inclined buttresses might be spotted. In addition, there was no video to work from, since astronaut Alan Bean had inexplicably pointed the Intrepid's TV camera right at the sun almost immediately after the astronauts began their first Extra Vehicular Activity, burning it out. Given that he was specifically trained not to make such a mistake, I tend to wonder if maybe someone at NASA was worried about what might actually be visible in the TV images. Apollo 11 hadn't had that problem, since it was mostly a symbolic mission that landed in the middle of nowhere.

It wasn't until Hoagland began examining some early NASA promotional films that he struck gold. At the height of the Apollo

Apollo 12 image showing astronaut Alan Bean in front of glass like lunar structures beyond the horizon.

Program, when new lunar landing missions were coming every few months, NASA's Public Affairs Offices were busy churning out promotional films and press release photos. Their job was to try to communicate the on-going success of Apollo to the American people and the Congress through the press. Their primary tools in that era (remember, this was long before the Internet) were mainly newspapers and four-color, glossy magazines—like *Life* and *National Geographic*, as well as television. For the magazines, they provided high-quality still photographs, and for television and school classrooms they provided a series of short films focusing on each new mission as it successfully ended.

It was possible that this extraordinary time-pressure on NASA Public Affairs to get the word out might really have allowed the "real stuff"—photographic details of what the crews really saw and photographed upon the Moon to slip through the cracks. Just after Apollo 12, NASA had released just such 16mm film, called

"INTREPID"

00:15:54:10

Screen capture from NASA film Pinpoint for Science (left) and from Apollo 14 photo AS14-66-9301, showing the same angled, structural buttresses over the horizon from 2 different lunar landing sites. The Apollo 12 Lunar Module Intrepid is visible on the left.

Pinpoint for Science.

After having an NSSDC 16mm print of *Pinpoint for Science* transferred to video, screen grabs were made and enhanced from the resulting video in the computer. The enhanced 16mm frames unquestionably revealed more of the reflective, glass-like ruins over the horizon from the Apollo 12 landing site, and the same massive inclined buttresses, slanting down beyond the lunar horizon in the distance. Bright star-like objects sparkled amid the amazing, steeply-pitched, stair-stepped lunar ruins, still attached to the visible but shattered geometric framework of the once vast lunar scaffolding that stretched miles overhead.

Given that two Apollo missions, 14 and now 12, had photographed the *same* crystalline geometry from two different locations and using two different photographic methods (70mm photographs and 16mm film), it's safe to say that we've found the ground truth we are looking for. But there is so much more...

The Russian Connection

It's not unreasonable to ask at this point why the Russians haven't ever chimed in on the subject. While they never actually put a man on the Moon, they certainly sent plenty of unmanned probes and those probes carried cameras, just as the NASA missions did. Unfortunately, much of the data gathered by a variety of Soviet space probes was (and still is) inexplicably off-limits to researchers from the West. Even after the Cold War officially ended, the flow of data from Russia did not expand, and at one point a private researcher was told in no uncertain terms to stop looking for lunar data from the Soviet era.

However, over the years a few choice items have emerged in the West. Back in the mid 1960's, the Soviet's had created the Zond series of spacecraft which were designed to conduct flyby's of various planets, including the Moon, Venus and Mars. Even by the standards of the time, the Zond probes were somewhat primitive, lacking the ability to place themselves in orbit around their intended targets and carrying cameras that were well below the quality of the instruments on American probes like the Lunar Orbiter series. Zond 1 and Zond 2 were initially successful, being sent to Venus and Mars respectively. It was thought in the West

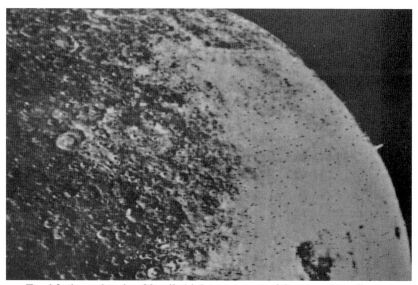

Zond 3 photo showing 20-mile high tower west of Oceanus Procellarum.

81

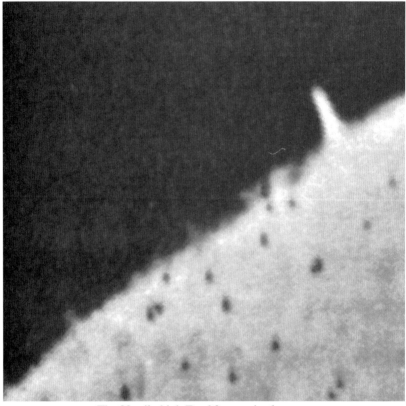

The 20 mile-high Zond 3 tower in close-up.

that Zond 3 was intended as a companion mission to Zond 2, and that perhaps the two spacecraft were intended to meet up in orbit around Mars. However, for reasons unknown, Zond 3 missed its launch window in 1964 and was instead launched in 1965 on a Mars trajectory, even though Mars was no longer in the same location. As a result, it was useful only as a guidance/telemetry test vehicle, and its only real data would be acquired in the flyby of the Moon that was needed to place it on a Mars orbit trajectory.

As it passed by the Moon with a closest approach distance of some 5,716 miles, the Zond 3 spacecraft snapped 23 photos and took 3 spectral images of the lunar far side. Centered mostly over the Mare Orientale impact basin, most of the images were fairly non-descript. But two stood out as completely remarkable. The first image (frame 25) was first published in defense contractor TRW's *Solar System Log* magazine in 1967. It shows yet *another* tower, sticking straight up from the lunar surface along the visible limb of the Moon.

This remarkable object is actually anchored somewhere over the horizon, near the towering structures on the western edge of Oceanus Procellarum that were visible from the ground by Apollo's 12 and 14. Again, by definition such an object (which is at least 20 miles high) *has* to be artificial because no natural object could be standing upright against the incessant meteoric rain of the last 4.5 billion years of the Moon's existence.

Close-up enhancements of the image shows that not only does the Zond 3 tower appear to be anchored at a point over the horizon, there are some odd and very geometric looking objects next to it. Again, a natural looking lunar horizon should be very smooth, not broken up as the area around the tower is, and of course nothing like the tower should be there at all...

The next Zond 3 image, taken thirty-four seconds later, had no tower visible at all, indicating it had, by that time, slipped over the lunar horizon due to the fast moving Soviet spacecraft's' motion and direction. After the tower vanished out of view, it was immediately replaced by an equally anomalous feature, found on image frame 28.

The second amazing Zond 3 shot was originally found in an official NASA publication, *Exploring Space with a Camera* (NASA SP-168, 1968), but is now available in high-resolution on the internet.

Located on the lunar horizon approximately a thousand miles further to the south, a large, eroded dome-like structure was plainly visible in the lower right corner of the image. Again, this "Zond 3 dome" extended several miles above the airless lunar horizon against black space. And, like the earlier "Zond tower," neatly aligned with the local vertical.

A close-up of this second Zond 3 dome reveals a significant amount of deterioration, undoubtedly due to long term exposure to the effects of meteor erosion. That said, the outline of a very geometric, structural building extending miles above the Moon is still clearly defined. Above the dome, the smaller remnants of

Zond 3 photo showing massive dome in lower right-hand corner.

what appears to be a lattice work type structure are visible under enhancement. Based on this, I suspect that what Zond 3 captured was in fact the battered remains of a watch crystal type of dome just beneath the larger protecting scaffolding we've seen other places on the lunar surface. Such an arrangement would provide an ideal engineering solution to the long-term issues involved with lunar habitation. Without an atmosphere to protect it from even the most mundane, everyday kinds of meteor strikes, a lunar base would stand little chance of surviving beyond a few years. But with a multilayered, miles-high scaffolding structure at the top and then smaller reinforcing "watch crystal" domes underneath, a lunar base might conceivably survive for hundreds of thousands if not millions of years.

But there is ground truth, and then there is ground truth. The next

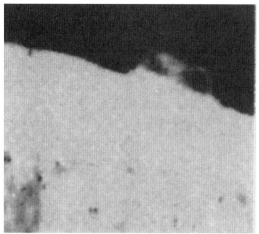

Close-up of the lunar "dome"—from Zond-3 (July 20, 1965).

obvious question is if these towering structures were all around the astronauts as they descended to and then explored the lunar surface, why didn't they *see* them? And if they did, why then did they not comment on them?

The first question is actually the easier one to answer. But the second one is the more revealing. After all, what if they did see these structures, but somehow just *forgot* that they did?

The True Colors of the Moon

If in fact there are these towering glass structures all over the Moon, scattering the light in geometric patterns all over the place, as in the Apollo 14 "Mitchell Under Glass" panorama, then that would go a long way to explaining the numerous reports of Transient Lunar Phenomena over the years. Instead of unlikely "outgassing" events, the observers were merely seeing specular reflections of specific panes of glass over the lunar surface. When the Moon moved a little bit, the phase angle of the reflected sunlight changes and the brilliant flashes of light go out. But this does not explain the odd changes in color that were observed repeatedly over the centuries. What could cause them?

Well, when light hits a glass surface, it tends, depending on the shape of the glass, to be bent and/or distorted. But, the glass doesn't stop the ray of light, although it might change its direction. The

85

effect is the same as if you shined light through a glass prism.

In addition to reflection, the other hallmark signature of glass is its ability for refraction—to bend light by slowing it down as the light passes through the glass. But light doesn't slow down (bend) uniformly when entering or leaving a transparent medium, like glass. It bends selectively, depending on the wavelength and the angle. This is the basic physics of a prism. Higher frequencies (shorter wavelengths, like blue and violet) are slowed down more. Thus, they are deflected through a greater angle of "dispersion" than lower (longer) wavelengths, like orange and red.

This results in the classic "rainbow" seen after light has passed through anything transparent, including glass, water droplets, ice, quartz crystals, or any refractive medium capable of differentially bending (slowing down the velocity of) light, compared to its velocity in a vacuum.

So if there really were massive amounts of transparent (glass) material on the surface of the Moon, the one *proof* of that hypothesis would be that the lunar surface below (and remnants of the structures above) should be all the colors of the rainbow, and all the colors in between. We should see not just red, green and blue, but pink and purple and yellow. But, didn't astronauts uniformly describe the Moon's surface as grey and colorless, a "magnificent desolation" of blandness? And didn't the pictures that were returned show the same thing?

Well, yes. But if you look at most photos of the astronauts on the lunar surface from that era, you'll also notice that the colors on their spacesuits are inexplicably faded. The fact is that most of these images have had the color toned down on them, to reflect the NASA position that the surface of the Moon is bland and nearly colorless. All you have to do is turn up the volume on the color, by increasing the saturation, and a whole different picture emerges.

Images, like the one below (NASA frame AS17-137-20990) show that not only is the lunar surface multi-colored, but those colors vary from place to place, just as they would if sunlight were being scattered by the towering glass structures. Instead of seeing a bland grey landscape, the Moon literally comes alive with pink

NASA frame AS17-137-20990. In the color version, orange soil in
the foreground is caused by iron oxidation in the soil, not from overhead prismatic effects. There are purple and pink rocks and mountains in the background of the color version of this photo.

mountains, blue rocks and purple boulders.

And the experiment was repeatable, using downloaded versions
of NASA own scans of their archived images. Everywhere we
looked, the Moon became a colorful, interesting place.

This also explains the earlier *Surveyor* images that came back
inexplicably colorful. Without knowing it, NASA had landed
beneath a virtual sea of glass prisms which turned the lunar surface
into a kaleidoscope of colors.

But again, this brings back the question of just how the
astronauts could have missed all this. The short answer is they
didn't. At least not one of them.

Alan LaVern Bean was the Lunar Module Pilot for the Apollo
12 mission and the fourth human being to walk on the surface of
the Moon. After returning from the Moon, he later flew on one

Self portrait of Alan Bean on the lunar surface. It has the exact color match for the Apollo 17 photos of yellows and reds.

of the Skylab missions and then retired from NASA in 1981 to devote his life to painting. Along the way, he made paintings of his experiences on the Moon with Pete Conrad on Apollo 12.

At first, these paintings depicted the lunar surface as we had all been told it was; bland and grey and colorless. But he then began to rethink his work, and as he did so, the images got more and more colorful until eventually they matched not only the colorful images we now see under enhancement, but the far-distant structures as well.

One of Bean's paintings, "Rock and Roll on the Ocean of

Storms," not only shows the distinct pink and blue colored lunar surface, it also shows lines in the sky that are eerily reminiscent of the "inclined buttress" structures seen in both the Apollo 14 photography and the Apollo 12 16mm film.

Now, these angled lines are actually the imprint of a moon boot that Bean puts on all of his paintings of the Moon, but there is no doubt that the placement of it and effect it has recalls the slanted structures seen in the Apollo images.

But these paintings all raise the question of just how well Bean truly remembers what he saw on the Moon. The fact that he was dissatisfied with the depictions that showed the lunar surface to look like the official NASA version is telling. His need to add color to the Moon's surface in his paintings ("I had to figure out a way to add color to the Moon without ruining it," he is quoted as saying) is also telling. Why? Maybe because there was something naggingly wrong with Moon as NASA chooses to depict it. Because of this, I think it is possible that Bean, like many of the other astronauts, has a hard time remembering what he saw and did on the Moon. There

"Rock and Roll on the Ocean of Storms" by Alan Bean.

have always been rumors that the astronauts were hypnotized during their technical debriefings to "help them remember." Given Bean's apparent struggle with getting his artistic depiction of the Moon "right," I tend to think this is likely.

I also take Bean's paintings as independent verification that the lunar dome hypothesis is correct, and that the Moon is far more colorful than we've been led to believe.

Which brings us to the first question; could the astronauts have simply missed seeing these towering lunar structures?

Again, the short answer is no. While the astronauts had filters on their helmets to supposedly help block out ultraviolet rays, these filters would most likely have helped, rather than hindered their ability to "see" the towering glass ruins in the distance. These ultraviolet filters were gold screens (similar to those used on fighter planes today) that could be slid down over the visors. These gold filters had the effect of actually *enhancing* light in the blue end of the visual spectrum, and we know that these enormous structures (due to a phenomenon called Rayleigh scattering) tend to shift light into exactly that same end of the color spectrum. So if anything, the gold visors would have made it easier to see the distant glass structures, not harder.

And there is one other piece of evidence that supports the idea that NASA had a pretty good idea about the lunar domes before they ever sent men to the Moon—the landing of Apollo 11.

By now, we've probably all seen the dramatic footage over and over again; the tense mission control engineers, the drama of Edwin "Buzz" Aldrin calling out the dwindling fuel resources, Neil Armstrong having to take over the controls manually to clear a field of boulders, and the scary "1202" alarm that no one could figure out. And then finally the dramatic "The Eagle has landed," call from Tranquility Base and the relief from "a bunch of guys about to blue" in Mission Control in Houston. But in reality, there is something far more telling about Eagle's dramatic descent to the lunar surface; what caused the "1202" alarm in the first place?

We now know, thanks to outlets like NASA TV and the History

Channel, that the 1202 alarm was a computer overload alarm. A modern cell phone has thousands of times the computing power that the Lunar Module did on that fateful day in 1969, so it might seem like overloading such a primitive computer was fairly easy to do. Not so. In fact, the 1202 computer alarm was so rare that it had not ever even come up NASA's landing simulations prior to the Apollo 11 landing. When it happened and the computer reset, Armstrong called back to Mission Control for an interpretation and was preparing to abort the landing if he didn't get the answer he wanted. In fact, according to Aldrin, he actually had his hand on the abort trigger while they waited for word from NASA as to the nature of the warning.

As it turns out, the reason for the 1202 alarm was that there were too many systems active on the Eagle and too much data was coming in. This was because Aldrin had done something he was not supposed to do; he turned on two different radars.

Unbeknownst to most people, the Lunar Module was equipped with two different radar systems, a landing radar and rendezvous radar. The landing radar was (obviously) downward looking and the rendezvous radar was side looking. According to the checklist, Aldrin was only supposed to activate the landing radar, but instead he activated both the landing radar and the rendezvous radar. As a result, the computer was overloaded with data and had to shut down and reset, hence the dramatic "1202" alarm.[1]

But the big question is why would he do this? If the Moon is exactly as NASA portrays it, then there is no need for anything but the landing radar. There are no towering glass structures to worry about hitting, so all you need to know is your altitude above the landing site. Since the Apollo 11 landing site was in virtually the middle of nowhere in the Sea of Tranquility, with no mountains around the flat plain for hundreds of miles, there was no need to worry about getting too close anything on the way down. But if Aldrin *knew* there were transparent, immense structures all over the Moon, then activating the side-looking rendezvous radar would have been a natural precaution.

So the next question is fairly obvious as well; if there are glass ruins on the Moon, where exactly is the rest of it located? As we'll see, it's all over the place. But one of the most interesting regions is someplace we never even landed—Mare Crisium.

[1] E-mail communication from Ken Johnston Jr. to the author. July 2, 2004.

CHAPTER FIVE
MARE CRISIUM

It is very strange the way the ejecta from Proclus
crosses Crisium. It is almost like flying above a haze
layer and looking down through the haze.
It looks like it is suspended *over* it.
– Al Worden, *Apollo 15 Command Module Pilot*

Mare Crisium (the "Sea of Crises") is dark colored, 350 mile wide "pond-effect" impact basin in the northern hemisphere of the Moon, just northeast of the Sea of Tranquility. Centered roughly 20° north and 60° east, it is near the eastern edge of the face of the Moon visible from Earth. It has relatively few major features, of which the major craters Proclus, Picard, Pierce and Cleomedes are the most prominent. The entire Crisium basin covers an area of about 65,000 square miles and is the source of one of the gravitational anomalies known as "mascons" (or mass concentrations) that I mentioned in the first chapter. No one knows what causes the mascons, but the most likely scenario is that there is a buried chunk of the asteroid or comet that created the mare basin beneath the surface. It has

a geometric appearance, a roughly hexagonal shape, and is the source of a number of enduring mysteries of lunar anomaly investigation.

The standard model for Crisium's smooth, dark appearance is that sometime after the major bombardment of the Moon which formed most of its plentiful craters, a large object struck the lunar

NASA photo AS15-M-0954.

Telescopic view of hexagonal Mare Crisium impact basin.

surface with such force that it made the area within the boundaries of Crisium molten. After some period of time, the area cooled and dark "basaltic lavas" (volcanic rocks) gave the region its dark, pond like appearance. Most of the visible craters in Mare Crisium are most probably smaller (and later) impacts that upset the relatively smooth surface of the "pond." This would seem to be supported at least in part by the presence of the mascon, which as stated above may be the remnants of the impact event which formed the region. None of this however explains how the region ended up shaped like a hexagon. It should have a rounded outline, like all other "normal" impact basins.

Mare Crisium is also the location of at least 12 Transient Lunar Phenomenon reports over the years. In 1882, an English mechanic who was also an amateur astronomer named J.G. Jackson made the following observation of Crisium:

Last evening (May 19th) on observing the moon's slender crescent, about two days old, I was struck with a very peculiar appearance on the westerly side of Mare Crisium, just on, or immediately within, the dark of the 'terminator.' It seemed a curved feathery mist or cloud lying just over the edge of the 'Mare,' and against the spur or range of mountains bounding the westerly side of the great valley. It seemed to be divided longitudinally by a faint dark line, and looked not unlike a feather. It must have been more than 100 miles long, by 40 or 50 miles wide. The definition was excellent, and I watched it for nearly an hour, showing it to several friends. In colour and appearance it was so strikingly different from the other illuminated parts, and so different from anything else I have ever seen on the moon, that I scarcely think it possible to be mistaken in its vapory character.[1]

In 1897, an astronomer named Jasper D. Hardy added to the lore in a report in an issue of the British Astronomical Associations *Journal*: "Now, at various times when studying the floor of Mare Crisium, I have noticed waves of light and shade, so that it was difficult at times to see objects with which I was perfectly familiar. Also clouds have passed over the object I have been viewing. These clouds have been seen by other observers, and are mentioned in Neison. That they belong to the moon there can be no doubt, and I gradually come to the conclusion that vapor of some kind still existed on the moon. I had my surmises verified on one particularly fine night (the best I ever had) when I plainly saw a well-defined cloud pass over the object I was copying."[2]

Another, more anonymous observation was reprinted in the science anthology *W.R. Corliss's Mysterious Universe; a handbook of astronomical anomalies:*

> Mr. Robert M. Adams has called to our attention a curious lunar observation by Mr. Robert Miles of Woodland, California, an observation rather reminiscent of Mr. Brian Warner's article on pages 130-131 of our November-December 1955 issue.
> Mr. Miles says in part:
> "I noticed a *flash* of white light that caught my eye. At first I thought it could have been a lunar meteor. But it kept flashing on and off... The light was very bright but changed its color to a very bright blue, like an arch light. It was brighter than the sunlit portions that I was looking

at."

Sketches indicate that the object in question lay on the night side of the terminator and perhaps about 100 miles east of the gap in the mountains on the east boundary of the Mare Crisium. Mr. Miles found the duration of visibility of the flashing light to be from 3:00 to 4:30 U.T. on January 17, 1956. The colongitude was then 320Â°.5 to 321Â°.3. These colongitudes seem inconsistent with the sketches, and the editor suspects that the U.T. date was really January 16.[3]

What each of these observations correlate, both in the transient manner of their observation and in the variance of color is that there seems to be historical evidence supporting the glass scaffolding theory over Mare Crisium. As we discussed in the previous chapter, light reflecting off of panes of glass and passing through multiple layers of glass would of course produce multi-colored specular reflections. Such light phenomena would obviously be transient in nature, ending when the Moon and Earth moved relative to the light source (the Sun). And the hazy, vaporous appearance is consistent with the observations made by Apollo 15 astronaut Al Worden at the beginning of this chapter. As panes of glass are illuminated, shattered remnants of the more beaten down edges of the structures will have a hazy appearance—like frosted glass— causing the illusion of being cloud-like.

But none of these less well known observations had the impact of a much more prominent anomaly in the Crisium region; the so-called "Mare Crisium bridge."

The first hint of controversy around the bridge was noted on July 30th, 1953 when a science writer for the *New York Herald Tribune*, John J. O'Neill, first spotted and reported observation of what he described as a natural "bridge" over two ridges in the Mare Crisium region near the crater Picard. Excited, O'Neill asked several prominent lunar researchers of the day to confirm his discovery.

Fairly quickly, a British astronomer named Dr. H. P. Wilkins

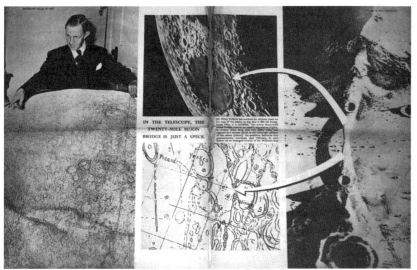

Coverage of the Mare Crisium "bridge" from *The Illustrated* in 1954.

did just that. He first reported his findings in a February 13th, 1954 issue of the Sunday tabloid *The Illustrated*.[4] He also published a more official paper on the bridge in the British Astronomical Association's *Journal* in February 1954 which included more sketches of his own observations. His characterizations at that time were of basically along two lines. First, he confirmed that he himself had seen what he thought was the same bridge across the two ridges that O'Neill had seen, and second, at least in this context, he asserted his belief that the "bridge" was a natural erosive formation on the Moon. Given that this presentation was in an official scientific journal, his caution is unsurprising.

Later (I'm unable to determine the exact date but it may have been December 23rd of 1953 or 1954), Wilkins stated in a BBC radio interview that he had confirmed a man-made (or alien-made, to be more accurate) "bridge" stretching across a significant portion of Mare Crisium. "It looks artificial. It's almost incredible that such a thing could have been formed in the first instance, or if it was formed, could have lasted during the ages in which the moon has been in existence," Wilkins was quoted as saying about the 12-mile long structure he sighted.[5]

Some subsequent reports have sought to downplay these comments by Wilkins, and to assert that he was quite

B.A.A. Journal]   [*Plate XI*

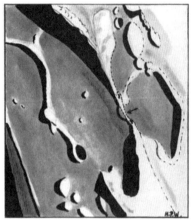

O'NEILL'S 'BRIDGE'
By H. P. Wilkins
1953 August 27

Illustration of the "Crisium Bridge" by H.P. Wilkins.

reserved about the possible artificial nature of the bridge that he and O'Neill both observed. But, a review of the transcript of the interview shows that is not the case and he was quite forceful about the probable artificial nature of the structure:

[Brian Forbes—Host] "Since the beginning of this Century, astronomers have been observing features on the surface of the Moon which have not been noticed before. During the last few years, many dome-like swellings have been seen through powerful modern telescopes. And only a few months ago, astronomers detected what is perhaps the most curious feature of all. It looks like a gigantic bridge. The Director of the British Astronomical Association— Dr. H.P. Wilkins—when interviewed, discussed this new discovery.

[Wilkins] If you look through the eyepiece, you will see one of the most interesting regions on the Moon... called the Mare Crisium. It's that comparatively small, dark oval marking.

[Forbes] Yes... I can see it now.

[Wilkins] I've mentioned this gap in the mountain barrier... but there now exists what looks like a bridge across this gap.

[Forbes] That's most extraordinary.

[Wilkins] Now this is a real bridge. Its span is about 20 miles from one side to the other. And it's probably at least 5,000 feet-or-so from the surface beneath.

[Forbes] It must be a most gigantic arch if it's 5,000 feet high.

[Wilkins] It certainly is.

[Forbes] How wide is it?

[Wilkins] The width is about a mile-and-a-half to 2 miles. It tapers narrows, rather—in the center.

[Forbes] Are you quite certain that you haven't mistaken it for some other object?

[Wilkins] Oh no, there's no mistake at all. It's been confirmed by other observers. It looks artificial. It's almost incredible that such a thing could have been formed in the first instance or—if it was formed—could have lasted during the ages in which the Moon has been in existence. You would have expected it either to be disintegrated by temperature variations or by meteor impact.

[Forbes] And when you say it looks 'artificial', what do you mean exactly by that?

[Wilkins] Well, it looks almost like an engineering job.

[Forbes] {exclamation of astonishment}

[Wilkins] Yes, it is more extraordinary.

[Forbes] And is it more-or-less 'regular' in outline?

[Wilkins] Absolutely regular in outline. That makes it all the more remarkable.

[Forbes] And does it cast a shadow?

[Wilkins] Yes, it casts a shadow under a low Sun. You can see the sunlight streaming in beneath it."[6]

The next day, brief cable reports on Wilkins' broadcast appeared in U.S. papers, but other than that it was apparently ignored. However, scientific sentiment against Wilkins quickly turned form curiosity to ridicule and he was eventually forced to resign from the British Astronomical Association because of his support of O'Neill's claims. He died a few years later in 1960, having never withdrawn his claims about the bridge in Mare Crisium.

Today, the general consensus is that both O'Neill and Wilkins were mistaken in their observations, the victims of an optical illusion that was never confirmed by other sources. This however ignores Wilkins own statements made in the BBC interview that "It's been confirmed by other observers." Meaning that, others besides himself and O'Neill had seen the structure too. Given the way in which Wilkins career was destroyed by his assertions, it's easy to see why these "other observers" have never come forward publically.

All of this serves as background for a number of more substantial discussions of other, far more interesting anomalies

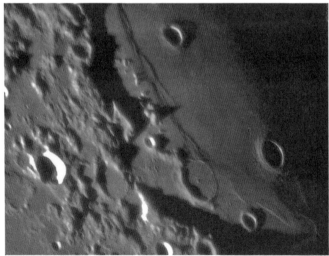

Area thought to be the Mare Crisium Bridge today.

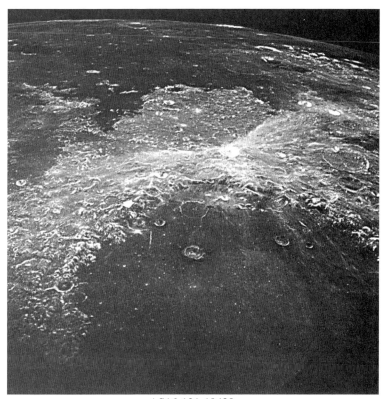

AS16-121-19438.

through the Crisium region. Most specifically those on NASA frame AS16-121-19438.

At first glance, AS16-121-19438 doesn't seem all that unusual or even interesting. Like a lot of shots of the Moon from orbit, it seems to show a fairly normal display the usual craters and highlands, dark maria and mountains. But then you start to notice the details.

For one thing, this Apollo 16 shot (the "AS16" in the frame number means it is an Apollo-16 photograph) contains a remarkable object right next to the crater Picard; a shimmering, vertical spire of glass very similar to the "Tower" in Sinus Medii. But there are also differences between the two objects.

First off, this "Crisium spire" is not only hundreds of miles it height, it is clearly (and obviously) supported by what appear to be guy wire like beams attached to it at several points along the

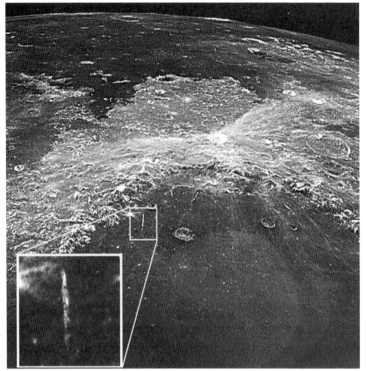

The "Crisium Spire" from AS16-121-19438 (inset).

vertical structure of the object. These luminescent support struts not only attach to the spire in logical places along its vertical axis, but they do so evenly on both sides. This symmetry is a tell-tale sign of engineering design intent, rather than some sort of optical illusion or natural phenomenon. The Crisium Spire also differs from the Shard and Tower in Sinus Medii in that is not, like them, anchored to a spot on the surface but rather appears to be suspended miles-high *above* the Mare Crisium plane, much like the Castle in Sinus Medii.

Further inspection in close up implies that the Spire is a corner post of what may be a large, box shaped structure stretching miles above the Mare Crisium plane. It looks very similar to the complex, box-like scaffolding seen in Sinus Medii in NASA frame AS10-32-4816.

But the goodies in this image of Mare Crisium hardly stop there. A closer inspection of the Spire shows that right next to it is the crater Picard, a distinctive feature in its own right. In most images,

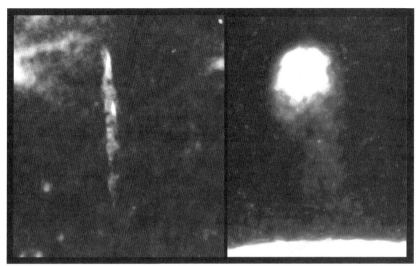

The Spire and the Tower side-by-side.

Picard looks for all the world like a perfectly normal impact crater (except for possibly the lack of a visible ejecta blanket). But in this image, it takes on the stark appearance of having a shattered glass dome partially covering it.

This unmistakable, pie-slice shaped structure glows in the sunlight, illuminating what must be the last remaining piece of a solid, watch-crystal like dome over the crater itself. Under intense enhancement, the wedge-shaped pie piece is even more obvious.

Also visible in the original enhancement are a series of parallel, dark striations along the northern rim of the crater itself. What

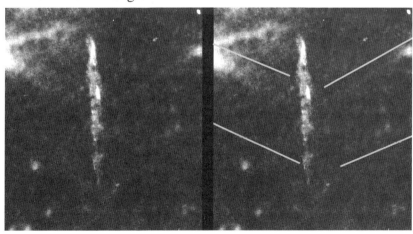

Support wires emanating from the Spire at 90° angles.

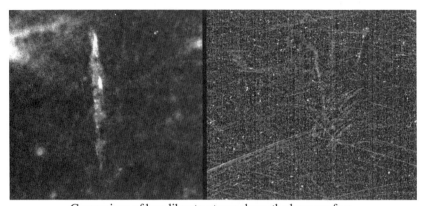

Comparison of box-like structures above the lunar surface.

these could be caused by is a bit of a geologic mystery.

Unfortunately, there are precious few other photos of Picard at the correct angle to provide comparison. One research group calling itself VGL (for Verified Gullible LUNAtics) has found one shot taken at an oblique angle—AS10-30-4421. In this image, Picard is seen from the side in a photo taken from a low angle by the Apollo 10 crew as the command module Charlie Brown flew past (not *over*) Mare Crisium in May, 1969. The first thing that is notable to me about the photo is a sort of gauzy haze which seems to be between the camera and the crater, much like Apollo 15 commander Al Worden's describes at the beginning of the chapter. According to VGL's analysis, they also noticed this odd haze. According to their report[7] "This could be an illusion caused by the

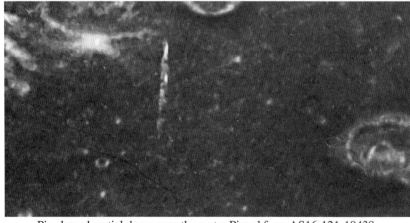

Pie-shaped partial dome over the crater Picard from AS16-121-19438.

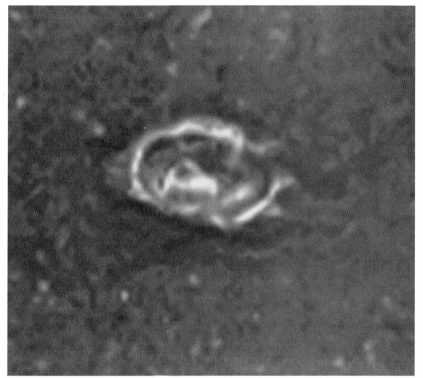

High contrast version of crater Picard from AS16-121-19438.

similarity of brightness and texture between the mare surface at the rim and the inner wall of the crater behind this indistinct section of the rim... It might, however, be a true obscuration of the rim by debris suspended above the mare surface between the camera and the rim of Picard."

While the standing, pie-shaped section is not visible in this image, close up enhancement of the area around the crater itself reveals what appear to be dark, structural arches rising over the crater rim. They would certainly account for the dark, striped markings seen on the rim of the crater from the orbital photo

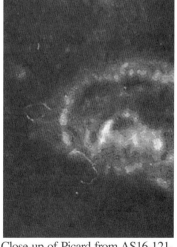

Close-up of Picard from AS16-121-19438 showing glass-like reflections and odd striations on northern lip.

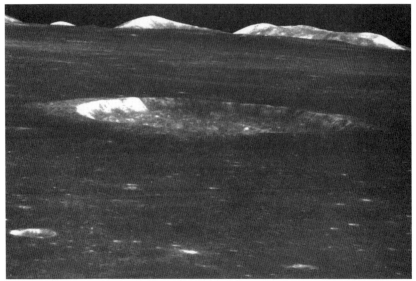

Picard crater from Apollo frame AS10-30-4421. Note the hazy appearance and the dark arches on the farside crater rim.

AS16-121-19438.

So how to explain these discrepancies? For one, it is not necessarily expected that we'd be able to clearly see a partially intact watch crystal type dome over a crater from every angle. Remember, these objects and the scaffolding are transparent, and they will reflect light only under ideal circumstances and when the camera is pointed at the reflection at just the right angle. Also, a glass structure such as this would need reinforcement, rebar if you will, to hold up the sections of pie shaped wedges to make such a dome. If these had collapsed, we'd expect them to fold over, in which case they'd tend to form arches just like the ones barely visible in AS10-30-4421. If only we could get a sharp, unobscured view of the crater, we might be able to see enough to resolve the question of

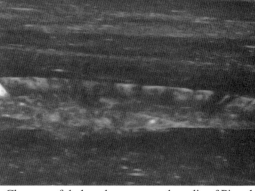

Close-up of dark arches over northern lip of Picard crater from AS10-30-4421.

106

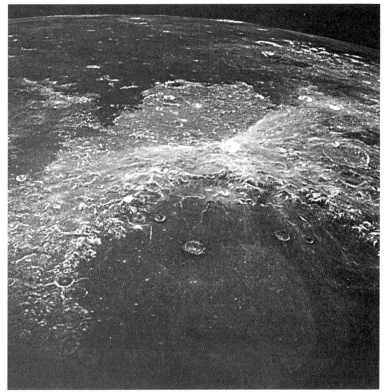

AS16-121-19438.

whether Picard once had a glass dome over it. But neither of the two images seem to be all that clear. The reason for this and the gauzy haze that astronaut Worden talked about is obvious; there is some type of nearly transparent medium between the camera and the crater in both images. And that medium is our immense scaffolding type crystal towers.

Let's take another look at NASA frame AS16-121-19438.

Besides the presence of the Spire, there are several other things "wrong" with this picture. At the bottom of the frame, you'll notice two oddly lighter hexagonal shaped areas of the photo over the lower part of Crisium. These are actually lens flares from the hand held 70mm Hasselblad cameras that the astronauts used and are fairly common on all the hand held photography. But once you get past that, there is still something not quite right about the lower half of this photograph which is not related to the Hasselblad lens flares.

If you look in the lower left hand corner, you can see a distinct

Enhanced version of AS16-121-19438. Note odd darkening in lower portion of image.

darkening, as if something were obscuring the craters and features below. This odd darkening spreads all across the lower portion of the photo and can only be accounted for by the presence of a semi-transparent intervening medium—like glass—being between the camera and the lunar surface below. This intervening medium—our crystal towers, are bending and obscuring the light just like an atmosphere would under the same circumstances. The result is a kind of "Levolor blinds" effect that is reducing the amount of light

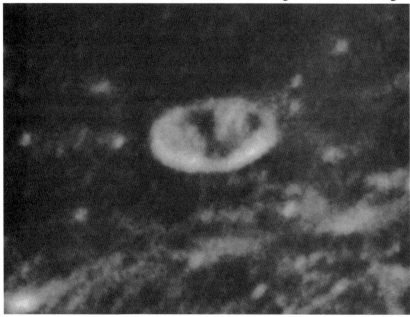

Satellite dish like crater Asada in the Mare Crisium region. Note the supporting structure under the "dish."

in the lower part of the image. Given the altitude of the spacecraft, these "blinds" must be truly immense, just as the images we've seen so far imply.

And there are other bizarre anomalies all over this image. As we look northward to the highlands, some of the most distinctive features, like the craters Asada and Proclus, actually look more like satellite dishes than craters.

Now, this could all be chalked up to an optical illusion, but upon further enhancement you can plainly see an engineering structure under the "crater" holding Asada up. It has vertical struts and horizontal members which appear to be the underlying support for the dish.

Notice also that there are regularly spaced dark areas under the dish where there are gaps in the girders holding the dish up. The dark shadow in the center of the dish is probably cast by the central mast of the dish subreflector. Unfortunately this cannot be made out at the resolution of the photograph. The point is, all radio

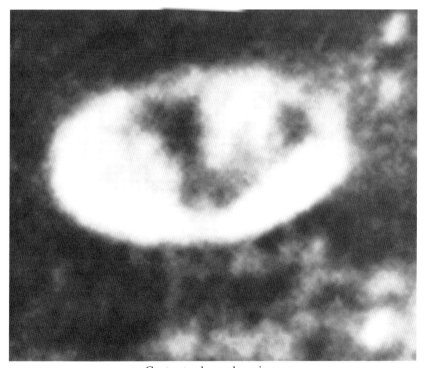

Contrast enhanced version.

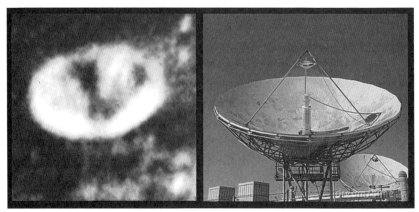

Supporting substructure of a radio telescope here on Earth.

telescopes and satelitte dishes have the same basic components; a parabolic reflector dish, a subreflector mounted on a mast over the dish, and an underlying subsctructure of griders and supports that not only hold up the dish, they also house the mechanism that rotates and aims the dish in a specific direction. Asada seems to have all of these distictive features.

Somewhat less distinct (because it is also somewhat farther away from the camera) but still impressive is the crater Proclus. Located just to the east of the Crisium basin at 16.1°N/46.8°E, Proclus is considered a fairly run-of-the-mill impact crater. Still, it has an unusual hexagonal shape and an extensive ray system emanating from it. Next to Aristarchus, it is the brightest crater on the Moon.

It is also one of the most colorful. Using techniques similar to the ones we discussed in the last chapter, amateur astronomers have captured the true colors of Proclus and the entire Crisium region quite successfully. Again, these color tone variations can only be explained by the presence of a glass-like medium bending and scattering light all over the lunar surface, creating multi-colored hues and all of the various tones in between. It implies that the entire Crisium basin lies beneath an extensive lattice work of glass structures.

But what makes Proclus stand out in AS16-121-19438 is its incredible resemblance—just like Asada—to a radio telescope or satellite dish. As we zoom up on it, even though the haze that astronaut Al Worden talked about, we can see that it has at least six

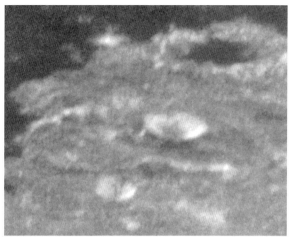

Crater Proclus from NASA frame AS16-121-19438.

regularly spaced support struts jutting out underneath it. Also visible are the regularly spaced dark gaps in between the struts, where it appears that access could be gained to whatever lies underneath the "crater." While it is true that very few images of Proclus show what we see here, this is the most oblique angle of the crater we have, giving us a rare side-on look at the formation.

And it looks like a satellite dish.

A quick side-by-side comparison of Asada and Proclus shows that we aren't simply seeing things. Both objects look exactly like you'd expect a radio telescope to look edge on and from a distance. Yet there are still some mysteries of the Mare Crisium region that remain unresolved.

In an effort to confirm these findings, I took to the web to find the latest high resolution scans of both NASA frames AS16-121-19438, and the more oblique AS10-30-4421. But what was

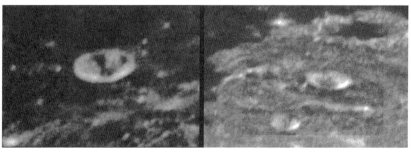

Asada and Proclus side-by-side.

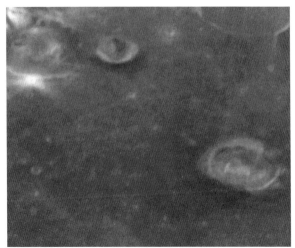

Official version of AS16-121-19438 with Spire
optically removed.

found only deepened the mystery. In doing a side-by-side comparison of the version of AS16-121-19438 that I had obtained through Richard C. Hoagland to the one I had downloaded from the NASA official sites, I began to notice some very disturbing differences. For one thing, the Spire wasn't there.

It was gone. Completely whitewashed. The wispy support wires are still somewhat visible, but the "Spire" itself has been completely and effectively removed. A quick comparison shows that in the officially released NASA version, the Spire simply ceases to exist.

Other features, like Asada and Proclus, seem to hold up better. They both retain their satellite dish characteristics and in fact some features can even be seen more clearly than in the official NASA

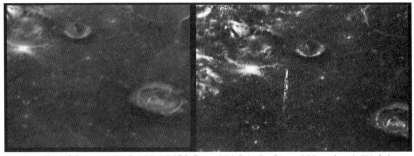

Apollo 16 frame AS16-121-19438 from NASA (Left) and Hoagland (Right).

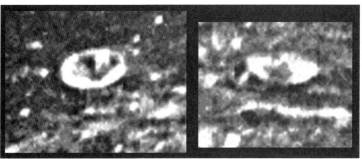

Craters Asada and Proclus from the Hoagland version of AS16-121-19438.

scans.

These discrepancies raise an interesting question. What if the Spire hadn't been digitally white washed from the official NASA version of frame AS16-121-19438? What if they weren't actually two different versions of the same photo, but rather two different photos altogether, both filed under the same official NASA frame number? Remember the 4822 Lunar Orbiter image of Sinus Medii? What if this was the same kind of skullduggery at work?

In order to clarify the possibilities here, a quick review of just where Richard C. Hoagland got his copy of AS16-121-19438 is worth repeating. As we told the story in *Dark Mission*, the image was obtained by a very lucky turn of events:

"The 'source' was well-known to NASA and (according to his account), 'regularly visited' at NASA Headquarters. He had been there, in the Administrator's office, a few days after Apollo 16 returned to Earth from its successful visit to 'the highlands of the Moon'—April 27, 1972. For some reason he was left alone in the Administrators office between meetings; he looked over and saw a massive bunch of Apollo photographs lying on the Administrator's desk. Bored, he casually leafed through a few... and was shocked by what he saw.

On impulse (he said later...) he quickly slipped one of the prints into his briefcase and—before the Administrator could return—abruptly left."

23 years later, this same anonymous source handed Hoagland the version of AS16-121-19438 which contains the Spire and numerous other anomalies. It is truly a sad state of affairs when a member of

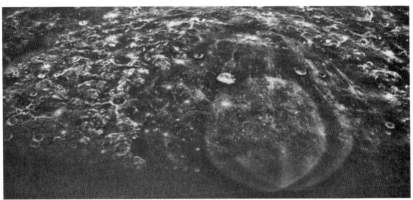

Lunacognita enhancement of the NASA version of AS16-121-19438.

the American public is forced to get the true story of what is on the Moon by swiping a picture off the NASA administrators' desk as opposed to simply having NASA reveal the truth themselves. But, such is the post Brookings world we live in.

That said, there is still quite a bit to be learned from the "official," sanitized version of the photograph. Some other intrepid anomaly hunters have been doing just that.

Although the official version of AS16-121-19438 is quite a bit blurrier and seems to have a significant amount of digital noise introduced into it, it is still useful for some research. A group calling itself Lunacognita has done some excellent enhancements of the photo and what it shows is not only revealing, but it is consistent with what we have already observed on the "Hoagland" version of the frame.

In all three of their enhancements, the obscuring medium in the lower part of the image is very clear, and it is obvious that there is something between the camera and the lunar landscape below, just as we see in the Hoagland and official versions. This all but confirms that the blurriness in both images is caused by something physical and is not a camera anomaly. In another enhancement, we can see the shattered remnants of this massive glass scaffolding high above the Crisium plane in the sky beyond. This indicates that the actual photo is shot down through a hole in the overall structure of the scaffolding, possibly punched there by thousands and thousands of impacts over the millennia.

(Lunacognita)

What these independent images show is that there is ample evidence for the theorized glass towers over Mare Crisium. We have several TLP observations over a couple of centuries at least, Earth based telescopic observations which show the multi-colored lunar surface as rays of light are bent, refracted and combined by the many panes of this glass structure. We have a NASA original photo—taken from the desk of the NASA administrator back in the 1960s—which shows an obscuring medium between the camera in orbit and the lunar surface below. We have a crystal clear image of a structural "Spire" supported by "guy wires" at 90-degree angles to it. We have a pie shaped portion of a former dome over the crater Picard. We have an oblique view of Picard that shows the dark, bent over supports of the dome that once encased the crater. And we have at least two craters which have underlying support structures and look more like satellite dishes than impact craters. We have an independent confirmation of a gauzy, haze-like blanket that one of the Apollo astronauts testified that he was looking down through to the Crisium basin below, which is entirely consistent with our model of a transparent superstructure over the Crisium basin. We have independent confirmation of the towering glass structures over Mare Crisium against a sky that should be absolute black. Structures, by the way, that are completely consistent with what we've already confirmed are over Sinus Medii. Did I miss anything?

Oh yeah, I did. We've also got a building on a mountain on the north shore of Mare Crisium and a couple of fully intact glass domes over craters there.

The North Shore of Mare Crisium

I first met Steve Troy in July of 1998 at a lecture given by Richard Hoagland in Phoenix. Given that I was computer savvy and Steve at the time was not, Hoagland suggested that we start working together to find and confirm evidence of lunar anomalies that might be consistent with what we had already uncovered. Steve had been quietly working to confirm Hoagland's artificial lunar dome hypothesis for almost 2 years and I had already put up my old Lunar Anomalies web site, so it seemed like a natural fit. Steve's approach was entirely analog, sifting through the reams of data and catalogs provided by a variety of NASA archives. Steve had been studying lunar anomalies this way since 1994, with an astronomical and geologic interest in the Moon for several years prior to that. He found numerous oddities and discrepancies in his studies, both with regard to the unnatural "geology" and the also with the photo's themselves. I provided image enhancement and confirmation of Steve's findings using standard digital techniques, and he likewise examined photographic negatives by eye of strange objects I had found.

One of the first frames that came up was the now familiar AS10-30-4421, the oblique image of Picard that had already been examined in some detail by Lan Fleming on the VGL website.

Fleming had focused primarily on the craters Pierce and Picard in Mare Crisium, and more specifically on the set of "arches" apparently over Picard on that image. While Steve and I had noted the arches, we were focused on another part of 4421.

The most noticeable thing right away is the very bright reflection in the upper right portion of the photo, right along the "shore" of the Sea of Crises. It is the brightest thing in the image by several orders of magnitude. That this bright reflection also caught the astronaut's eyes is very obvious. In fact, it appears that the astronaut who was holding the 70mm Hasselblad camera took this photo to specifically to capture the bright flash on film.

AS10-30-4421 is part of a power winder sequence taken

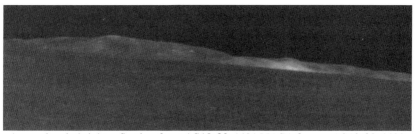

Anomalously bright reflection from AS10-30-4421 NASA frames AS10-30-4420 and 4421.

by one of the Apollo 10 astronauts as they passed just south of Mare Crisium at orbital velocity (about 3,500 Miles per Hour) with a clear view across the dark plane. Frames 4414 through 4420 were taken through one of the optically perfect windows on the spacecraft and in putting them together, you can track the progress of spacecraft as it passed by just to the south of Crisium. Essentially, the astronaut holding the camera didn't move at all, he just kept clicking frame after frame. You can see this plainly by the changing position of Picard in each image. By frame 4419, Picard has disappeared from view off camera right. Then on frame 4421, the astronaut suddenly turns the camera back along the path that he'd already photographed and snaps another photo. The obvious question is why, and the obvious answer is the sudden flare of light appearing along the mountain range to the right.

Obviously curious as to what had caught the astronauts' attention to the degree that he felt compelled to turn back and get a shot of the flare, Steve made several sectional enlargements of

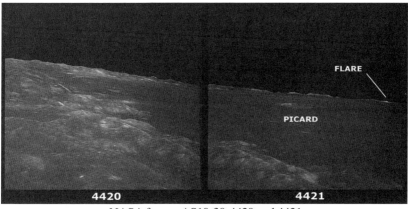

NASA frames AS10-30-4420 and 4421.

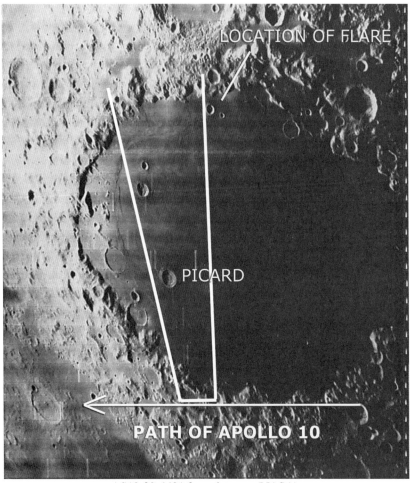

LOCATION OF FLARE

PICARD

PATH OF APOLLO 10

AS10-30-4421 footprint map (NASA).

the area and sent them to me for enhancement.

The first issue was to determine where exactly the flare had come from. According to Steve (and the NASA footprint map) Apollo 10 was flying almost due west and looking almost due north across Crisium, and the edge of the photo was just west of the craters Cleomedes F and Cleomedes Fa. If this was correct, it seemed we were looking at a flare of some kind up on the mountains themselves. Even as bright as it was, it might not be anything but a very flat, reflective face of a very non-descript mountainside. Then I got my hands on the sectional.

What I saw, unmistakably to my cye at least, was a

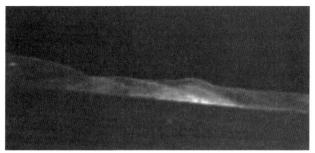

Close-up of glass domes over Cleomedes F and Cleomedes Fa.

stunning edge-on photo of a clear glass dome. You can see the mountain range behind the dome and follow its contours. There also appeared to be something geometric actually *inside* the dome. It also seemed to have 2 distinct "corners" or edges, rather than blending in with the mountains behind.

This meant that the dome itself stands apart from the mountain chain, which implies that the dome is sitting on the Mare Crisium plane *in front* of the mountain range beyond. The other thing I immediately noticed was that the brightest area was on the right side of the dome, where there seemed to be a second dome visible behind and through the first (and larger) one.

This second dome is more opaque from the perspective of 4421 than is the main dome, and the area where it overlaps the main dome is the area of brightest albedo. This would be explained by the light sources of the two domes adding their brightness as they reflected the sunlight into the camera.

Under enhancement, the less distinct left edge of the Main Dome became more clearly defined, and this helped to confirm its location in front of the mountain chain along the North shore. There were also a series of interesting details that emerged as I got deeper into the data.

The geometric area that appeared to be inside the dome has some very interesting features, to say the least. The first was an odd bowling pin like

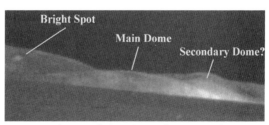

Annotated image of glass domes from AS10-30-4421.

119

Contrast enhanced.

appendage that seemed to be protruding from within the dome itself. And on the hillside above (and beyond) the dome was a tube-like, geometric formation I called the Phoenix.

It is constructed of 2 cylindrical objects interlocking with the vertical cylinder bending over to the left, giving the impression of a bird with wings outstretched in flight. I have no idea what this object might be but it stands out dramatically from the drab mountain chain it rests on and bears no resemblance to any explainable lunar geology. Its reflectivity suggests a metallic or crystalline construction. This is reinforced by the fact that such a rounded surface would not reflect this much light if were made of typical highland material.

The "Bowling Pin" is a dark bulbous object just to the right of the "Phoenix" which actually seems to protrude through the main glass dome. This appears to be an antenna of watchtower of some kind, and its middle portion is definitely obscured by the glass like material of the "Dome" itself. Note how the upper tip of the Bowling Pin is encased in the same glass like material and blots out the upper rim of the dome behind it. Again, a spike like object such as this has no place in standard lunar geology, and I am at a complete loss to explain it as anything but artificial.

Obviously, I wanted to follow this up and see if I could find

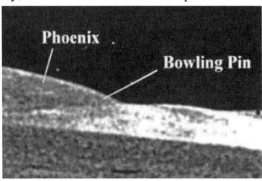

The "bowling pin."

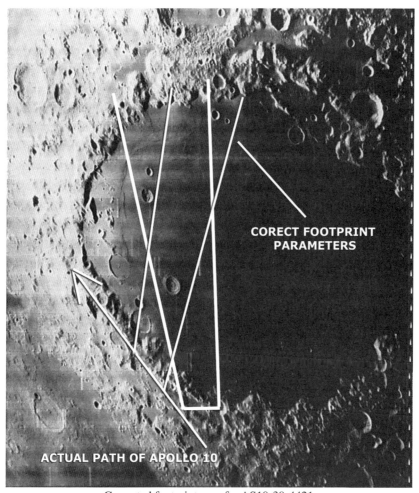

Corrected footprint map for AS10-30-4421.

another image of the area where the twin domes might be located. But looking at a number of images of Crisium yielded no results. But then it occurred to me; what if I was simply looking in the wrong place?

In reviewing the NASA footprint map and comparing it to the AS10-30-4414 to 4421 series, I realized that the footprint map couldn't be right. If it was, and the spacecraft was travelling due west along the south side of Crisium, then the crater Picard *had* to get smaller and smaller in each of the successive photos, because it would be getting farther and farther away. But it didn't. Picard, if anything, got *bigger* in each successive photo.

Lunar domes, imagined and real.

What this meant was that the original image map was in error (or deliberately altered) and Apollo 10 had actually been traversing the edge of Mare Crisium on a northwest trajectory. It also meant that two features which were supposedly outside the photographic parameters of 4421 were actually in full view of it—and they fit all the requirements to be the source of the bright flare and the watch crystal like domes which caused it. So this solved the mystery of why we couldn't find what might have caused the reflections off of the mountains. The reflections were off of two domes on floor of Mare Crisium itself, covering the craters Cleomedes F and Cleomedes Fa.

So here we had the first photographic evidence of at least one (and possibly two) intact, watch crystal type glass domes over a pair of craters in the Mare Crisium region. I have to emphasize how lucky we are to even have this confirming image of the glass domes over the Cleomedes craters. If it weren't for the quick reaction the astronaut holding the camera to turn back and shoot the picture under ideal (and rare) circumstances, we might never have had a clear, edge on shot of the domes. And we might never have had a clue about what is waiting for us on the floor of Mare Crisium if we ever successfully land there.

These types of domes would have represented the last line of defense against meteor bombardment for an ancient alien settlement there. With the eroded, structural lattice type structures overhead, providing the upper layers of protection, the lower domes would have enabled a pressurized, livable environment on the lunar surface. What might still be preserved underneath those structures is a matter of conjecture because we don't know if they are still sealed and pressurized, or if they are long since

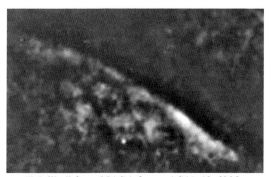

"Malibu" from NASA frame AS11-42-6223.

abandoned. Sadly, there are precious few photos of the inside of Cleomedes F and Cleomedes Fa, and they are of rather poor quality, so a full analysis of the contents of these lunar "arcologies" (architectural ecologies) will have to wait for further data from later missions. But I have to wonder if we get the data if we'll even be allowed to see it.

As part of our study of the mountains around the north shore of Mare Crisium, Steve came across a number of oddly geometric formations along the mountain tops. One was especially intriguing and very close to the locations of the two glass domes. Spotted on NASA frame AS11-42-6223, one hilltop showed what appeared to be a structure built right into the mountainside.

Nicknamed "Malibu" by Steve because of its location on a hillside, this building conformed to the contour of the hillside, had straight square walls at 90-degree angles to each other, and also had internal walls (again at 90 degrees) dividing the structure into numerous subsections or rooms. There was also a central spine visible and what might have even been doorways and entrances in several spots.

Now, on the negatives ordered by Steve in the mid 1990s, Malibu stands out like a sore thumb on the image. But today, if you download AS11-42-6223 from the online NASA archives, all you will see is a faint blur on the hillside where the structure appears. Again, it appears that Brookings is still alive and well, and NASA is still whitewashing images to hide the "real" Moon.

No manned landings were ever attempted on Mare Crisium, al-

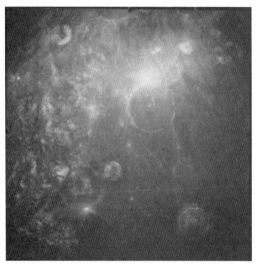

Al Worden's "gauzy haze" over Mare Crisium.

though one unmanned mission, the Soviet *Luna 24*, did land there successfully far from any of the structures we've talked about in this chapter and return a soil sample to Earth. Still, there have been enough solved and unsolved mysteries uncovered in the area to conclude that it is a major source of Ancient Alien ruins, if not active archeological sites. But there is one issue with Crisium that is still outstanding; why is it so dark?

As I said earlier, the traditional model of the formation of the maria, the dark areas on the front face of the Moon, is that they contain dark basaltic lavas. But samples returned from Crisium and other mare show very little of the volcanic rock that could account for this. As we have seen from Apollo 15 Command Module pilot Al Worden's own comments, there seems to be a gauzy haze over large parts of Mare Crisium, and this haze can be confirmed in numerous photos of the area.

The working theory is that this gauzy haze is caused by looking down through an intervening medium, a transparent (or semitransparent) lattice-work of glass structures that were erected in the long distant past as a meteor shield to protect the surface installations much like our own atmosphere protects us here on Earth. When the sunlight hits this just right, you get a bright flare, like in the image above. But in general, when looking at the Moon from a distance, the

Earthrise.

maria like Crisium tend to be very dark. There's a reason for this, and it all but confirms the existence of massive glass domes.

When we look at the Earth from space, we see white clouds, light colored land masses and very deep blue oceans. The reason for this is that the clouds are the highest in the atmosphere, meaning that they are reflecting more light back to the camera, and at a faster rate. Since they are returning more light, the clouds are the lightest. The surface areas meanwhile are darker, because they are a bit farther away from the camera than the clouds and therefore the light has to travel farther before it is reflected back. The deep blue oceans are therefore the darkest, because the light has to travel all the way to the ocean floor before it is reflected back to the camera's "eye."

The same idea, but slightly inverted, applies to the Moon.

The lighter areas are the areas where these glass structures most likely do not exist or are at least less densely packed. Therefore they reflect the light back more directly. The much darker maria are the areas where the glass scaffolding acts like an atmosphere, scattering the light into the blue range of spectrum and making it pass through multiple layers (and multiple prism's) before it finally reaches the

The Moon. Darker areas may represent locations of highest density of glass ruins.

surface and is reflected back in the multi-colored spectrum we see in today's color enhanced images of the Moon.

What this all adds up to is that the Moon is a very different place than the long-dead, rock strewn desolate landscape we've been led to believe by NASA and the Brookings crowd. What you are about to find out is just how different.

But first...

[1] *W.R.Corliss's Mysterious Universe*, page 237 (Jackson, J.G.; English Mechanic, 35:326, 1882)

[2] *W.R.Corliss's Mysterious Universe*, page 236 (Hardy, Jas. D.; British Astronomical Association, Journal, 7:139-141, 1897)

[3] *W.R.Corliss's Mysterious Universe*, page 227 (Anonymous, Strolling Astronomer, 10:20, 1956)

[4] http://the-moon.wikispaces.com/O'Neill's+Bridge

[5] BBC radio interview (Patrick Moore's Armchair Astronomy, Patrick Moore, 1984, Patrick Stephens Ltd., U.K

[6] Transcript by Isabel L. Davis, of Civilian Saucer Intelligence, N.Y.C. "The Flying Saucer Conspiracy," Maj.(ret) Donald E. Keyhoe, Henry Holt & Co., First Edition 1955

[7] http://www.vgl.org/webfiles/lan/picard.htm

CHAPTER 6

WHO MOURNS FOR APOLLO?

In the last decade, I have become increasingly alarmed as a particularly silly and damaging urban myth has begun to take hold. Promoted by a few well known authors such as David Percy, Bill Kaysing, Ralph Rene' and the late James Collier, this latest twist on the current conspiracy nation fad is based on a simple, if unbelievably naive and absurd notion—that the Apollo missions and subsequent Moon landings were faked. Even after Percy's late 1990's *Fortean Times* article was pretty much taken apart by readers, references to the "fake" landings began to creep into popular culture, springing up in such diverse places as Jay Leno's Tonight Show monologue and commercials featuring ESPN's Chris Berman. I think it is very important, especially for young readers who may not have studied this in great depth, to understand the difference between a good conspiracy theory and a bad one. And the idea that the Moon landings were faked is definitely a bad one.

Admittedly, I thought this whole issue was put quite nicely to rest in August 1997, when my co-author on *Dark Mission*, Richard C. Hoagland, debated James Collier on Art Bell's *Coast to Coast AM* radio program. The results of that debate can only be described as an unmitigated humiliation for Collier, who turned out to be totally out of his element and misinformed on the general subjects of space travel, physics, engineering, NASA, and Apollo itself. But I have to keep in mind that every ten years or so, a new generation comes along and they have to be reminded of the truth and the facts.

I have been accused, along with Mr. Hoagland, of defending NASA on this count only because I need a legitimate manned lunar program to support my Ancient Alien theory. I hope this chapter will make it clear there simply is no logical basis for the faked Moon landing conspiracy theory.

Let me be clear; I am unabashedly a conspiracy theorist. I am 100% convinced that there has been a cover up by NASA of some extraordinary discoveries made in the course of the agency's 40-year year history, and I think the data we've already covered in this book proves that. That said, one thing they did not do – unquestionably—was fake the Moon landings. In fact, most of the charges made, not just by Collier and Percy, but by others who have picked up the mantle of their assertions, are so absurd, so easily discredited, so lacking in any kind of scientific analysis and just plain common sense that they give legitimate conspiracy theories—like mine—a bad name. Frankly, I suspect that may ultimately be the point of this whole thing after all.

Almost from the moment that Neil Armstrong and Buzz Aldrin set foot upon the Moon at Tranquility Base, the rumors began circulating that the whole thing was faked. I have always felt that there was something a little more to this than simple stupidity or naïveté, something a bit insidious about the whole thing. That was more than confirmed in the Forward to *Dark Mission*, when Richard related his memories of being handed a pamphlet claiming the landings were faked even before Neil and Buzz had splashed down on their return trip from the first lunar landing. What made that moment so extraordinary was not that someone had made up a pamphlet making such a claim, it was that the person who authored it was being escorted around the NASA press room by a NASA press officer to make sure every reporter got one.

So yeah, the rumor that the Moon landings were faked was actually started with the able assistance of NASA itself more than 4 decades ago.

I assumed that as time went by, the notion would weaken and falter, rather than gain momentum as it has recently. I have come to wonder, given my own stance on the whole question of what the Moon program was really about and what the astronauts really found, if there wasn't perhaps something a bit "conspiratorial" about the promotion of this patently absurd and demonstrably false conspiracy theory. Richard's story, recounted as I said in more detail in *Dark Mission*, all but confirms that.

As you will see, some of the specific citations of evidence that the landings were faked can actually be more easily explained not just by a complete rejection of the Moon Hoax theory, but a combination of conventional explanations and the glass ruins model of the Moon. The way light scatters on the Lunar surface; the size of solar reflections in the visors of the astronauts (which are way out of proportion to their counterparts on modern day Space Shuttle missions), the sometimes secretive stance taken by the astronauts and the Agency; the very peculiar qualities of the film in the cameras taken to the Moon by the astronauts, all point to something bigger and more interesting then we have been led to believe by NASA itself. In this chapter, I will try to sort out the most common claims being made, highlight the rebuttal evidence, and show that the Moon landings were something quite extraordinary after all.

Yes Virginia, We Really Went to the Moon

There are three major thrusts used by the fake landing advocates to bolster their claim: First, that the radiation exposure suffered by the astronauts was not survivable; second, that the photographic evidence "proves" that the landings were staged in a Disney movie studio somewhere; and third, that the mechanical aspects of the mission—the pure mechanics and physics of the journey—are not as claimed and therefore must be faked.

Each of these claims is based on misinterpretations, misrepresentations, or just plain ignorance of the realities of space travel. It is not a coincidence that many of the believers in this myth are too young to actually remember the Moon landings. If they had been old enough to watch the missions on live TV, they would have known that most of these claims are nonsense. For this article, we will deal with each of these claims in separate sections, and try to directly address the key sub-claims being made.

Section One – The Photographic Evidence

Percy is one of the primary drivers of this particular set of claims, but Collier and others have added to it. Let's list a few of the claims one-by-one and address them.

Claim 1 - The shadows don't fall right in images taken on the lunar surface, proving that there are multiple light sources, like professional stage lighting using high- powered lamps. Since the Moon has only one light source, the Sun, these images (these people claim) "have to have been shot on a sound stage somewhere."

This one is usually based on images like the one above (taken from an Apollo 17 TV transmission), that seem to show the shadows of the astronauts coming from different lighting sources. However, a logical approach to this problem reveals that there is nothing at all mysterious about either the shadows or the light sources. If, in fact, the shadows were cast by different light sources, each astronaut would have two shadows instead of just the one each we see here. Yet, in the images that the "Moon Hoaxers" cite, there is consistently only one shadow being cast, indicating that the Sun is (as it should be) the single and dominant light source.

So, how to explain the seemingly divergent shadows in this image? If you look closely, you will see that the astronaut on the right is on a slight rise above the astronaut on the left. This has not only the effect of lengthening his shadow, but also if the slope is greater in one direction, say to the left of the astronaut on the right, it will tend to flow and elongate in that direction.

It's important to keep in mind that the Moon has a very rough and uneven surface, with lots of slopes, rises and potholes. As a result, many of the shadows will appear to be non-parallel.

Shadows cast in divergent directions from Apollo 17 TV broadcast.

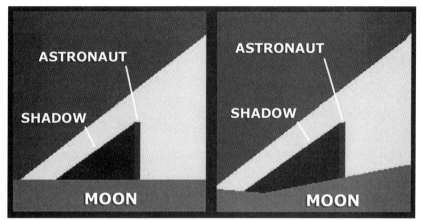

The geometry of shadows.

In a sense, the Moon Hoax advocates are correct here; there is no comparison to be made from lunar landscapes and terrestrial ones. But, it is because the surface of the Moon is so uneven, not because there are multiple light sources, i.e. lamps, casting the "wrong" shadows.

Also at issue is the photographic equipment used by the astronauts on the lunar surface. Shortened wide-angle lenses, like the ones on the hand-held Hasseblad 70mm cameras used by the astronauts, will distort otherwise parallel shadows. Simply pull some outdoor photos from your own personal collection and see for yourself.

Claim 2 - The foreground of many images of the astronauts on the Moon are filled in with light, while the shadows remain absolutely black, again proving that there are multiple light sources.

In this one, the argument is that with his back to the sun, the astronaut's suit should be as dark as his own shadow stretching out in front of him (see Apollo 16 image, above). Since there is no light diffusion in an absolute vacuum, NASA "must" therefore have used reflectors or fill-in lamps to illuminate the astronaut for this photograph. The truth is, there *is* evidence of a "reflector" in this image—but it's the lunar surface itself.

Obviously, the lunar surface is a fairly bright gray color. It

131

is known, from the Apollo samples brought back and analyzed in Houston, to contain a LOT of glass beads and metals like aluminum and titanium, with a lot of other reflective and refractive minerals in it. All of these materials tend to kick light directly back toward the source of illumination with very high efficiency. This is one reason why the Full Moon is so much brighter (than other phases) in the night sky; the sun is "behind" the Earth. The effect of the sunlight hitting the lunar surface and being reflected back toward the sun itself creates a backscatter that fills in the astronaut's bright white shadowed suit with an excellent fill-light effect. And the fact that the shadow is so dark on the ground in front of him is proof of exactly the opposite of the claim being made by the "Moon Hoaxer" crowd. It shows that indeed, the astronaut is standing upright in a harsh vacuum, where his suit can "see" the illumination from the surrounding lunar landscape. By stark contrast, almost no light at

Light scattering off the lunar surface.

all has seeped into the shadow, because it's lying flat on the ground and cannot "see" anything but black space overhead. It is, as it should be, extremely dark and sharp.

Interestingly, as to the question of multiple light sources, some of the leading debunkers of the Moon Hoax theory, like Dr. Phil Plait, have also made a very significant mistake. It is flat wrong, as many of them have stated, that the Earth is a "very significant" light source on the Moon. When full, the Earth is on the order of 68.4 times brighter than a full Moon as seen from Earth. It also takes up something like 13.5 times as much sky. But, that's not the whole story.

The Earth is—maximum—100 times the brightness of a Full Moon (I'm going to overestimate a bit to prove the point). The apparent magnitude (brightness) of a Full Moon is about -13. The equivalent magnitude of the Sun is about -27. Subtracting, that's a difference of 14 magnitudes. Since each 5 orders of magnitude correspond to a factor of 100 in brightness, a difference of 14 magnitudes corresponds to almost 100 X 100 X 100, or a factor of a million. Allowing for the ~100 times greater reflected brightness of the Earth (at "Full Earth") the direct lunar sunlight is still at least 100,000 times brighter than the Earth's illumination.

There is no way that the slide films used by the crews could have detected that feeble amount of "Earthlight" on the lunar surface, even in the shadows, because the exposures were set for the full, sunlit view.

Of course, we have our own thoughts on this. Some of the debunkers must be realizing that backscatter is insufficient to account for some of what we are seeing on the lunar surface photography. To come up with an explanation, they have resorted to the (obviously incorrect) "Earth light" angle. But of course, it's really more interesting than that...

Claim 3 - There are no stars in the background from pictures taken on the Moon.

This one keeps coming up, but the answer, while obvious, is somewhat complicated by our own lunar conspiracy theory.

Usually, Moon Hoax advocates cite any number of pictures of the lunar surface showing an absolute black background, but this one above of John Young saluting the flag in front of the LM Orion is quite prevalent. Anyone with the slightest knowledge of photography can easily put this one to rest.

On an airless body like the Moon, any brightly lit foreground object must be photographed with a very short exposure time. Otherwise, the image will be badly overexposed. Any background pinpoint light sources—like, say, stars that are literally trillions of miles further away—will not show up at all. Likewise, if the photographer wants to capture the background stars, he is going to have to use a very long exposure time, which means that the foreground will be totally washed out in one blob of overexposed

Apollo 16 astronaut John Young saluting the US flag on the surface of the Moon.

134

light. Obviously, there would be no real benefit to taking such an image, since the point of the lunar surface photography was to document the activities of the astronauts on the lunar surface—not to stargaze.

Even so, there are plenty of pictures from missions like the Surveyors which prove my point. In order to get the stars even to show up (for navigation and location purposes), the Surveyor spacecraft cameras had to use (in one example) a three-minute time exposure to record them. By contrast, the average exposure time of the hand-held, film photographs taken on the lunar surface by Apollo astronauts is about 1/250th of a second—or an average of forty-five thousand times shorter than the exposure required to actually record stars in the airless lunar sky. Clearly, if these "disbelievers" believe at least in the reality of the unmanned lunar landings (which at least some say they do), this simple example should satisfy even the densest Moon Hoax advocates as to the nagging question of why stars aren't visible in the background of any Apollo surface photographs. Because they're simply too dim.

This whole process is complicated by the fact that in a

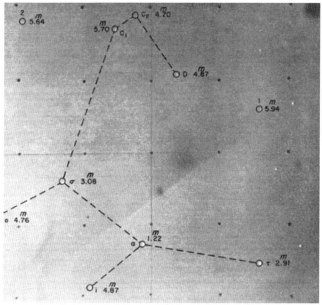

Surveyor 6 photo of the constellation Scorpius taken from the lunar surface.

135

vacuum, the problem is made even worse, the light far more intense, and the exposure must be even shorter. The Moon Hoax advocates also seem to have forgotten that they are basing most of their "analysis" on press release photos, which are invariably cleaned up before release to the press. So of course, these sanitized press kit images would reflect what we all would expect to see, an absolute black background.

So contrary to what the Moon Hoax advocates have been saying, the sky above the astronauts *should* be absolutely black. And in fact, on most of the prints that they have been looking at, web based images, press release photos, and even new prints from the archives, it is. The problem is that while the sky should be absolute black, and does appear that way in images presented by the Moon Hoax advocates, it most demonstrably is not absolute

Four NASA images showing glass structures over Occanus Procellarum.

Close-up of NASA color image AS16-120-19187 showing blue-shifted glass structures obscuring view of Earthrise.

NASA color image AS16-120-19187, showing glass-like intervening medium obscuring the lunar landscape.

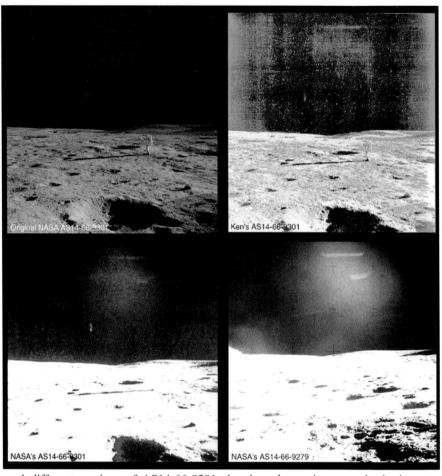

4 different versions of AS14-66-9301 showing glass ruins over the horizon. (NASA)

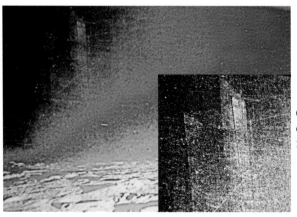

Geometric glass structures over Sinus Medii from NASA image AS10-32-4816.

AS14-66-9301 The "Mitchell Under Glass" image from the Ken Johnston Collection.

Apollo 12 image showing astronaut Alan Bean in front of glass like lunar structures beyond the horizon.

Self portrait of astronaut Alan Bean.

Imagined lunar dome as it might have appeared eons ago.

"Rock and Roll on the Ocean of Storms" by Alan Bean.

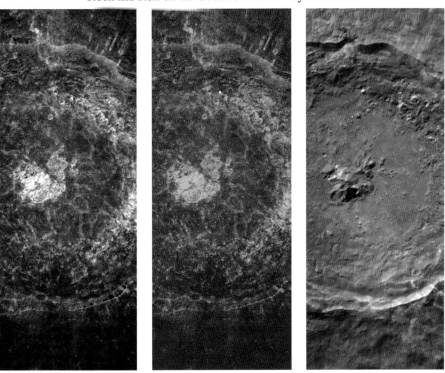

Clementine composite image of Tycho crater.

Apollo 16 astronaut John Young saluting the American flag on the surface of the Moon.

Enhanced color image of the crater Proclus from Earth-based telescope. Color variations and intensity are caused by bending of light through immense, glass-like lunar structures suspended over Mare Crisium region.

Color enhanced image of Lunar Rover and Shorty crater. Note Orange soil, pink mountains and blue rock outcroppings. Colors are the result of prismatic light scattering from overhead glass ruins.

4-frame composite of 'Data's Head' from Jack Schmitt's panorama of Shorty crater. Red stripe appears to be painted on metallic surface of the artifact.

Orange soil near Shorty Crater. Note lavender rocks nearby and pink mountains in background. (NASA)

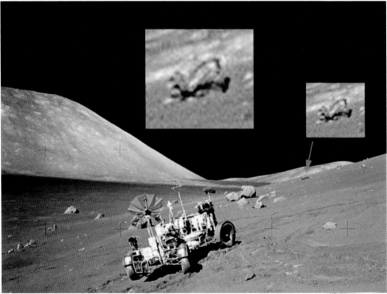

Image of mechanical device near the Lunar Rover from AS17-140-21409.

Apollo 16 astronaut John Young.

black in the images we examined earlier in this book.

What Ken Johnston's 1ˢᵗ generation prints showed was quite another story—that the sky above the astronauts was far from blank—it was in fact filled with a strange, bluish, geometric set of ruins. So the problem is exactly the opposite of how it is stated by the Moon Hoax advocates. The sky should be black, but it isn't.

One amusing sidelight of this famous Apollo 16 photograph is that it is used on several web sites as "proof" that many of the pictures taken on the Moon are fake, since John Young "... is casting no shadow at all" on the lunar surface. In fact, all it really shows is how dumb most of the Moon Hoax advocates really are. If you actually look at the picture, you will see that Young is casting a shadow to the right side of the picture a few feet away. How can this be? Why is the shadow not "attached" to young's feet? Well, because in this famous sequence, John Young is leaping into the air as he is saluting, while fellow astronaut Charley Duke snaps the photo. Many Moon Hoax advocates, too young to have actually watched this all on live television, look at this picture and mistakenly believe that Young is standing on the slight dome shaped rise in the background, when in fact he is in midair (well, OK, *mid-vacuum*). This famous sequence is also a good way to show that the astronauts are indeed in the one-sixth gravity of the Moon, since in order to get this kind of elevation on Earth (especially with the bulky, several-hundred-pound spacesuit and backpack on), Young would have to have the leaping ability of Lebron James. There are many video sources available today which show this famous live TV sequence.

Claim 4 - In some images, a huge light source can be seen reflected in the astronaut's visors. This has to be a very bright, nearby source.

This argument is essentially a variation of the first argument. Occasional images, like the ones above (taken from the Apollo 17 EVA TV transmissions and Apollo 14), seem to show a very bright, huge light source taking up almost 25% of the astronauts visor. Moon Hoax advocates argue that this is proof of a large light source (a stage flood or a spot, again) positioned very close to the astronauts. What they are missing here is essentially the same geometric problem they missed with the "bent shadows" argument. The gold-covered helmet visors that the astronauts wore were very convex shapes—similar to automotive wide- angle side mirrors included on many current models. Like the surface shadows in the earlier images above, this curved helmet has the effect of severely distorting the reflections, making them appear much smaller (and thus farther away) than they actually are.

The problem is the sun in the visor reflection pictured here appears much larger (and therefore closer) than it possibly could. The explanation for this is firmly grounded in the theory we've already covered; the presence of ancient, glass-like ruins on the Moon. It is

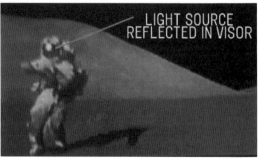

Image frame capture from the Apollo17 mission.

138

these ruins, sticking up above the lunar horizon and physically intervening between the low-angle sun and the Apollo astronauts roaming across the surface, which create the magnified halo of scattered light seen in the gold visors. Since this area of forward scattering is much larger than the optical size of the sun itself, it makes the reflection appear disproportionately larger.

Claim 5 - There are no views of the Earth in pictures taken from the surface of the Moon.

This one also is just plain wrong. Collier was among the most enthusiastic promoters of this mistaken notion, based on studying only a few press release photographs from NASA. Below is an Apollo 17 photo of a large boulder, with the Earth in the background, taken by an astronaut with a hand-held Hasselblad 70mm

camera There are dozens of other such examples. Since all the non-hand-held pictures taken on or in orbit around the Moon were using a media other than 70mm transparency film, these photos had to have been taken by a human being—an Apollo astronaut—physically present either on the lunar surface or in space around the Moon.

Apollo hand-held photo of the Earth from the surface of the Moon.

Claim 6 - How could NASA take TV images of the LM ascending on Apollo 15, 16, and 17 if there was no one on the Lunar surface to man the camera?

Now, most of the Moon Hoax accusations are pretty dumb, but this one really has to take the cake. As you can see from the collection of images above (from two different missions) on the later Apollo missions (15-17) the astronauts left the TV camera

Still frame captures of liftoff from the surface of the Moon.

pointed at the LM so that viewers on Earth could watch the liftoff. Initially, the camera was unable to track the ascent stage as it rose into space, but by Apollo 17, NASA had figured a way to get the camera to track upward and follow the spacecraft. All they did was calculate the time difference for radio transmissions from the Earth to the Moon and send a command for the camera to pan upward to follow Lunar Module ascent stage as it rose. So the answer to this one is also simple and obvious—the camera was remotely controlled from Earth.

Section Two – The Mechanical Arguments

Most of these claims come from James Collier's book *Was it only a Paper Moon?*

Claim 7 - The astronauts could not have egressed and ingressed the LM because they could not fit through the hatch and there was no room to even open the hatch in the LM.

It's hard to know just how to respond to this one beyond simply stating that it is wrong. As you can see from the artist's concept below, the astronauts were positioned on either side of the central cockpit panel, with the main EVA door between them. There was in fact plenty of room

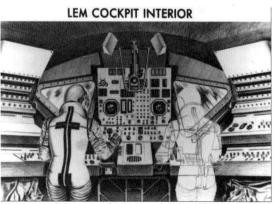

Diagram of LEM cockpit interior.

140

to open the hatch. On Apollo 11, Armstrong would have been manning the left position in this view and Aldrin the right. The door was latched to Aldrin's side, necessitating that the door be swung open inward, and effectively "trapping" Aldrin momentarily on his side of the LM. In fact, this is the main reason that Armstrong went out first. Once he was out, Aldrin was able to close the hatch, move over to Armstrong's position, and exit the LM himself.

Buzz Aldrin exiting the LM.

As to the issue of whether the astronauts could fit through the hatch, as you can see, they must have. This is a photo taken from a film shot by Armstrong of Aldrin egressing the Lunar Module. The entire sequence is available from the NASA archives, and shows the whole procedure from start to finish, including Aldrin opening the hatch and crawling through it.

Furthermore, if it turns out that the astronauts could not fit through the hatch, this will come as quite a shock to our friend and contributor Ken Johnston, Jr. He spent many hours in the vacuum chamber at Houston, fully suited up including the backpack, crawling in and out of the full scale mockup of the LM, to test exactly that. He'll be very upset to learn that he wasted all that sweat for nothing.

Claim 8 - The Lunar Rover was too big to fit in the LM.

Well, this is strictly true if you take the measurements of the Rover when it was fully deployed and assembled. However, the Rover came packed into a very tight little package which fit neatly into the space provided in the LM.

To deploy the Rover, all the astronauts had to do was pull on two nylon cords and the it popped right out of its berth and down to the lunar

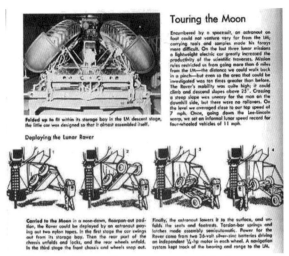

Diagram of Lunar Rover stowage and deployment.

surface. As it did so, the wheels, which were folded over (as you can see in the photograph above) deployed outward and were then locked into position. The main purveyor of this claim that the Rover was too big to fit into the LM is Collier, who took his measurements by going to the Johnson Space Center (where there is a full scale mockup of the Rover in its deployed configuration) and then compared those numbers to the containment bay on the LM. Anybody with a rudimentary knowledge of engineering could have figured this one out — simply by looking at the hinges which allowed the wheels to fold out when deployed. This whole aspect of the controversy could have been avoided by a trip to the film archives which have plenty of footage from the Apollo missions showing the astronauts actually unfolding and deploying the Rover on the Moon.

Claim 9 - The astronauts could not get from the Command module to the Lunar module with their space suits and pack on.

Again, strictly true. They never did go from the CSM to the LM and back with their packs on for one very good reason — they didn't have to. The back packs were stored in the Lunar Module the whole time. Beyond that, they did not wear their packs at all until they actually went out on their moon walks.

As a further side note on this, Aldrin has told the story in recent years that during the process of putting on their backpacks for the first Moon walk, one of them inadvertently broke off the arming toggle switch for the ascent stage's main engine. Had Aldrin not used a ball point pen he was carrying to flip the switch to the armed position, he and Armstrong would have been stranded and died on the lunar surface.

Claim 10 - There can't be any pictures taken on the Moon because the film would melt in the 250° temperatures.

Any normal film exposed to 250° would indeed melt at that temperature. There are only two problems with this Moon Hoax claim—this was no ordinary off the shelf film, and it was never exposed to those kinds of temperatures in the cameras.

The 70mm film used in the Hasselblad cameras the astronauts carried was a very special transparency film designed specifically (under a NASA contract) for hostile environments like the Moon. According to Peter Vimislik at Kodak, the film would at worst begin to soften at 200° F, and would not melt until it reached at least 500° F. So, a worst case scenario of 250-280° F for a totally uninsulated, non-reflective camera would still be well within the film's operational parameters. The film itself, in terms of its light- gathering abilities, was also quite amazing. It was a special extended range color slide film called "XRC" that allowed the astronauts to take perfect National Geographic quality pictures on the lunar surface, even though they were hardly experienced photographers. This has truly opened up whole web pages of controversy—with the Moon Hoaxers claiming that such a film simply doesn't exist In fact, Richard Hoagland actually used many rolls of this super lunar film, back when he was advising Walter Cronkite at CBS.

In addition, the cameras were also protected inside a special case designed to keep them cool. Although it is true that in the direct, airless sunlight the temperature can reach upwards of 250° - 280° Fahrenheit because there is no air, it's also fairly easy to keep cool for the same reason. The situation is a lot different than in your oven, for instance. With no convection or conduction, the only type of heat that is of concern on the Moon is radiative. The best way to reflect radiative heat is to wrap the object (like a camera or person) in layers designed to reflect as much heat as possible, usually by simply being colored white. Most all of the astronaut's clothing and the camera casing were indeed white, which very efficiently directed heat away from the both the astronauts and camera film.

Claim 11 - The LM engine was very powerful. How come it did not leave a crater below the spacecraft? Why didn't it kick up any dust when it landed?

The truth here is once again very straight-forward. At all of the landing sites, the astro-nauts found that the lunar surface had about a two inch layer of dust. Below that was pretty much hard pan. As you can see from the image below from

Blast pattern from Lunar Module descent engine.

Apollo 11, not only is the upper layer of dust blown away in a radial pattern (as if from a thruster?) there is also a small depression below the nozzle. Since the LM descent engine only made about 3,000 pounds of thrust (compared to a modern jet fighter which makes between 18,000 and 22,000 pounds of thrust), this is pretty much as any engineer or geologist would expect things to look.

And what of the charge that no dust was kicked up by the LM as it descended? Again, we'd recommend any of the fine NASA vid-eos on the Apollo program. They show that in each and every case, the LM did indeed create a literally blinding swirl of dust blown radially outward from under the descending LM, as it groped its way down, balanced on its 3000-lb thrust engine, to its final lunar resting place. You simply have to be willing to find the films and watch them.

Section Three – The Radiation Arguments

Claim 12 - The astronauts could not have survived the trip because of exposure to radiation from the Van Allen belts and solar flares.

Actually, of all the issues put forth by the Moon Hoax advocates, this is the one that requires the most digging into. The Van Allen radiation belts are a pair of toroidal (donut-shaped) belts of high-energy electrons and ions trapped in the Earth's magnetic field. Any object leaving Earth's orbit to visit the Moon or beyond must pass through them. The inner region is centered at about 1,865 miles above Earth and has a thickness of about 3,100 miles. The outer region is centered at about 9,300 – 12,500 miles above the surface of the Earth and has a thickness of between 3,700 – 6,200 miles. According to a document called "Radiation Plan for the Apollo Lunar Mission,"[1] the radiation in the belts was of *some* concern to the scientists working on the problem. However, they actually considered a rogue solar flare to be a much bigger problem.

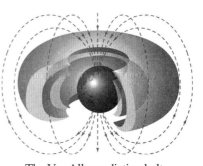

The Van Allen radiation belts.

In fact, as stated in this official government report, the scientists working on the problem of Van Allen radiation considered it to be minor compared to other design hurdles to be conquered. Protection against the radiations of the Van Allen belts was a complex problem recognized long before Apollo or even before the advent of manned space flight. Prior to 1958, scientists knew that ions and electrons could be trapped within Earth's magnetic field. 1957 and 1958 were designated as the "International Geophysical Year"—a time in which the first artificial satellites were launched by both America and the Soviet Union for the first overall surveys of the Earth from space. The Soviet's Sputnik and America's Explorer I (the latter instrumented by James Van Allen) were both launched in

145

1957, and 1958, respectively. Explorer I carried Van Allen's Geiger counters to observe cosmic rays, but the instruments mysteriously appeared only to work at the lower altitudes of its elliptical orbit. Explorer III followed two months later with more sophisticated instruments, and detected very high levels of radiation. Vast numbers of energetic particles were detected hitting the counters at higher altitudes, and in specific, belt shaped regions. These "belts" (which had literally saturated Explorer I's more limited detectors, accounting for their apparent failure to detect the belts at higher altitudes) were eventually recognized as doughnut shaped regions where both protons and electrons are trapped within Earth's magnetic field. Particles within the belts were seen to spiral around the Earth's magnetic lines of force, therefore changing orientation continuously in relation to a moving spacecraft.

These two radiation belts have different origins and compositions. The inner belt discovered by Van Allen occupies a region above the equator, is a byproduct of high-energy cosmic radiation, and is populated by protons of energies in the 10-100MeV (Million electron Volt) range. These can penetrate spacecraft and on prolonged exposure, damage instruments and astronauts. The outer belt is an electron-plasma trapped in the magnetosphere from the Sun's expanding solar wind, and has energies in the 0.1–10 MeV range.

Before we proceed, it is necessary to define a few terms. A RAD or Radiation Absorbed Dose, is a unit of measurement that determines the actual absorbed amount of radiation by any given material. The material can be plastic, metal, or biological, or anything else for that matter. It does not define the degree of biological damage that can occur to the absorbing individual, since different types of radiation can cause differing levels of damage to human tissues. Rather, it is a blanket number defining total radiation exposure of all types.

The REM, or Roentgen Equivalent Man, is a unit used to derive a quantity called an equivalent dose. This relates the absorbed dose in human tissue to the effective biological damage of the radiation. Not all radiation has the same biological effect, even for the same amount of absorbed dose. To determine equivalent dose (REM), you multiply absorbed dose (rad) by a quality factor (Q) that is unique to the type of incident radiation.

There are several mitigating factors that can affect the amount of damage done to human tissue by radiation exposure. Even if a specific type of radiation is very damaging to humans, if you limit the time that a person is bombarded by that radiation, you can reduce the effect on the person's cells. This is why fair skinned people will not get sunburned if they are only outside without a sunblock for a few minutes. If they are outside for a few hours, they can get a very painful radiation burn. Continuous exposure of this type of radiation over years can lead to skin cancers.

This time element, or exposure, must always be considered alongside the intensity and quality of the radiation a person is exposed to. As a rule, acceptable doses for high risk individuals like astronauts are expressed in RAD's. For example, 100 RAD's will induce vomiting, over 150 RAD's are fatal if untreated, and a 500-rad dose is fatal even with medical treatment. Delayed effects include cancer and other genetic changes. These long-term effects can occur even when the dose rates are far below the thresholds for any prompt effects.

After years of extensive study, NASA's solution was simple; avoid exposure to the radiation in the belts by keeping the spacecraft at low Earth orbit altitudes while in parking orbits, and then send them through the belts at high speed. The eventual escape speed, some 25,000 miles per hour, would have passed them through the belts in less than an hour, keeping their dose well below 1 RAD. There was a modicum of shielding from the equipment, but in the end this was not necessary as the transition speed kept the dose below harmful limits—both going to and returning from the Moon.

As to the issue of solar flares and the danger they presented, there simply weren't any major flares during any of the Apollo missions. So the biggest reason that none of the astronauts died from their radiation exposure was that the actual doses, in RAD's, that the astronauts received were quite small. NASA spent millions to develop the necessary technology to insure that the astronauts that went to the Moon were protected from the physical threats of deep space and they were monitored at all times while travelling to and from the Moon.

Claims from "Conspiracy Theory—Did we Land on the Moon?"

In 2001, Fox Television broadcast a TV special called "Conspiracy Theory—Did we Land on the Moon?" Because this program raised some additional issues we did not specifically cover yet, in the interest of closure I have decided to address them here. Make no mistake, I was so unimpressed with this laughably stupid presentation initially that I was quite willing to let the previous part of this book be my final statement on the matter. But I guess I just can't resist.

Claim 13 - There are cross hairs on pictures taken on the Moon that appear to be behind objects, rather than in front of them, where they should be.

The crosshairs, called reseau marks, were geometric indicators specifically put in the Apollo cameras by the vacuum deposition of a set of whisker-thin aluminum "crosses" on an optical glass plate, subsequently placed just in front of the film plane. The purpose of this (according to NASA) was to enable the NASA-Houston developers of the film to align multiple image panoramas vertically and horizontally, so that they might appear geometrically correct when printed.

The Fox special showed four examples of the crosshairs appearing behind objects in the pictures. One example each from Apollo 11 and 16, and two from the same frame on Apollo 12. In addition, I found another example on the Project Apollo

Reseau crosshair appearing behind an instrument.

image archive, AS16-117-18818. The four that were presented on the show are the same ones that seem to make the rounds of all the Moon Hoax sites, and I have not seen any other examples although, as I just demonstrated, it seems fairly easy to do so.

Crosshair blending into astronauts' white suit.

The argument made by the Moon Hoax advocates (primarily the late James Collier, David Percy, Bill Kaysing, "brilliant lay physicist" Ralph Rene, and the late Dr. Brian O'Leary) is that these obscured reseau marks "prove" that the photos taken on the Moon are faked. They imply that the marks were added *after* the photos were taken to make it appear that they were taken on the moon but that NASA screwed up some of these fake reseau marks.

It's hard to follow this convoluted logic. If NASA were faking these pictures in a movie studio at Area 51, as Fox and Kaysing alleged, why wouldn't they simply have used cameras with the same aluminized, pre-marked plates in them that were used on the real Apollo cameras? Wouldn't that be easier than painstakingly adding the marks one by one by hand to every single Apollo hand held photograph? And if the pictures were all faked, why add the marks at all? Wouldn't it be easier to just avoid the whole hassle by skipping the reseau marks completely?

Now, in fairness, some of the Hoax crowd has claimed that these apparent retouches aren't truly just stupid mistakes by NASA after all, but a deliberate code. They claimed (without evidence) that certain "patriotic Americans" working in the NASA photo lab and outraged by the huge hoax being perpetuated by Apollo, deliberately made "little mistakes" in placing the crosses on some photographs. The purported purpose was to telegraph the fact that the whole Moon program was as fake as the photographs themselves. As ingenious as this explanation might appear to some, there is a far simpler and more likely solution.

For one thing, in all the pictures presented, the marks are obscured by white areas of the pictures. Be they the white stripes of the American flag, the white covering of a scientific instrument, or an astronauts' spacesuit. Anyone who has ever developed color film will tell you that white tends to bleed a bit into other colors, and given that the crosshairs are only few thousands of an inch across, it's easy to assume that this is the explanation. As far as I know, none of the Moon Hoax advocates has ever actually examined the negatives of these frames, either. Certainly, if the blotting out of the crosshairs is an anomaly of the printing process, then the negatives should probably have the full reseau marks visible and we will have our explanation. It is also probable that the highly reflective white surfaces just got slightly overexposed in some photographs, simply blotting out the razor thin marks.

But, failing in that, there is another, even better explanation. The pictures *were* deliberately altered.

Wait a minute; doesn't that imply just what the Moon Hoax advocates are saying? That the photos are really are faked—and for over thirty years someone's been trying to blow the whistle?

No. Of course it doesn't.

There is a huge, huge difference between "altered" and "faked." It's a fairly safe bet that numerous Apollo pictures were altered, and there is nothing sinister at all about it.

In each case that I have seen, including the one that I personally found, the altered crosshairs (if that's what they are) are from the NASA press release collection for the Apollo Program. It is an entirely common practice for press release photos to be "cleaned up" before publication, and there is no reason at all to think that the Apollo Missions to the Moon were excluded from this practice. It is difficult for us, in this digital age, to appreciate the importance of print media in the days of the Apollo Program. Many Americans got their view of Apollo not from TV, but from newspapers and magazines like *Look*, *Life*, and *National Geographic*. A potential press release quality photo (say, of an astronaut standing next to the American flag) would of course be cropped, blots and flaws would be airbrushed out, and perhaps an offending reseau mark would be kept from crossing in front of the Flag itself.

One final point: Over the last 15 years, Richard Hoagland, Steve Troy and I have examined literally thousands of Apollo pictures taken on the lunar surface. The only ones which have had this issue with the crosshairs have come from the press release collection—exclusively. We have simply not seen this phenomena on any of the non- press pictures. Or more importantly, their photographic negatives.

Claim 14 - Other photographic anomalies

Another picture that has made the rounds of the Moon Hoax sites is the infamous "C" rock. This is supposedly a picture from an Apollo surface photograph showing what appears to be the letter "C" etched on it. Moon Hoax advocates claim that this is a mark for stage props, so the stage hands would know where to place the fake rock. But, if this were a "prop rock" from a studio collection, why would NASA use a classification system (the Arabic alphabet) that limited them to only 26 "prop rocks" in any given lunar landscape? Obviously, they wouldn't.

At any rate, after finding the photo in question (AS16-17445-46), a cursory examination of the prints and negatives in the NASA master files revealed that the letter "C" was simply a hair or fiber that got on the scanning bed while the photo was being scanned for NASA archive websites. No analog or other digital source ever showed the "C" anywhere on the rock in question.

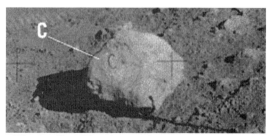

The Apollo "C" rock – a hair on the scanned print.

151

But enough of this. There are many additional claims made by the Moon Hoax advocates, all of which fall pretty much into the same category of silliness or nonsense. There is no question that the Apollo missions were real, and that NASA sent six teams of astronauts to the Moon to investigate its mysteries. The far more interesting questions are just what exactly they were looking for, and just what exactly they found...

[1] http://www.braeunig.us/space/69-19.htm

CHAPTER 7

APOLLO 17

In our previous book *Dark Mission*, Richard C. Hoagland and I discussed just what each of the Apollo missions was intended to *really* do behind the scenes as opposed to what was publically admitted as the scientific and technical goals of each mission: What the "Dark Mission" of each of the Apollo landings might be. Reading that book might help in understanding some of the ritualistic motives involved. In the end, we came to these loose conclusions about each of them:

Apollo 11—Ceremonial landing in the middle of nowhere, far away from glass ruins. Timing of the mission was designed around Buzz Aldrin's Masonic "consecration ceremony" which took place before the first Moon walk.

Apollo 12—Proof of ability to navigate through the towering glass ruins and land at a spot safely among them. Testing of TV equipment to ensure that glass ruins beyond horizon could not be seen on live broadcasts. Consecration of 2nd lunar Masonic temple at same location as Surveyor 3, which landed in almost identical spot on Hitler's birthday 2 years earlier.

Apollo 13—Ceremonial Sacrifice to the ancient "gods" of NASA, Isis and Osiris.

Apollo 14—Symbolic "resurrection" of Alan Shepard as "Osiris/Orion." Investigation of Cone crater. Multiple photos taken from ground level of same distant glass structures as seen in nearby Apollo 12 landing site photos and films. First surface use of new color TV cameras.

Apollo 15—Tight, pinpoint landing in highlands area up against Apennine Mountains. First deployment of Lunar Rover allowing for exploration of mountainous "arcologies." Orbital probes and spectrographs take close-up readings of compositional make-up of areas like Mare Crisium. Glass ruins likely mapped and material make up confirmed.

Apollo 16—Lunar Module "Orion/Osiris" lands on the Moon on April 20th, 1972 (Hitler's birthday). Belt stars of Orion are 33 degrees above landing site at touchdown. Investigation of nearby "mountains."

Apollo 17—Salvage mission.

Obviously, out of all the missions, Apollo 17 is the most intriguing. It appears to be nothing less than an investigation of lunar arcologies, massive artificial structures that have taken on the appearance of mountains after eons of lying in ruins.

Fortunately, two associates, Keith Laney and Steve Troy, have been looking at Apollo 17 data for some time. Between them, they had done extensive photographic studies of the Apollo 17 mission to Taurus-Littrow, obtaining many early generation negatives. Keith had even posted an extensive analysis of all of this on his website. But the question still remained; what was so special about Apollo 17 besides its historic place in the books as the last mission of the Apollo program? Why did NASA choose to land so far north after all the other Apollo landings had taken place near the equator?

The first thing that's notable about the Apollo 17 Mission is the very dangerous look of the landing site itself. Positioned at 19.5° N by 31° E, the target landing ellipse is in a narrow valley amongst the Taurus-Littrow highlands. This was by far the riskiest Apollo landing of them all, as Gene Cernan would be required to set the lunar module Challenger down among gigantic (6,500 to 8,200 feet tall) mountains on a valley floor littered with large craters. In order to even reach the Taurus-Littrow site, NASA had to abandon long-standing mission rules requiring "free-return" trajectories (a lunar orbit insertion trajectory which would allow the spacecraft to loop around the Moon and return to Earth in the event the Command and Service Module engine didn't fire), as well as prohibitions against launching at night. They even abandoned mission rules covering the roughness of landing sites in order to accommodate the desire to land at Taurus-Littrow. So what could be so fascinating or critically important about the mountains at Taurus-Littrow that would justify all these risks?

The most obvious clue was the ritual aspect of the 19.5° N latitude landing ellipse. As we extensively covered in *Dark Mission*, this "19.5°" number appears over and over in NASA literature and numbering conventions. Without going into detail, there appears to be something sacred to NASA about the numbers 19.5 and 33, especially when expressed as angles in a 360 degree measurement system. Given this well documented obsession, it might have made a certain sense to NASA to look for a possible landing site at this latitude on the assumption that their chances of finding a set of Ancient Alien artificial ruins there would be greatly enhanced.

In more closely examining the proposed landing site years later, it became obvious to Keith Laney what the attraction was. There, in almost the center of the landing ellipse, was a massive, hexagonal mountain. Officially dubbed the "South Massif" for purposes of navigation ("massif" is the French word for "mountain"), the mountain has at least four clearly visible and near-equal length sides, and the implications of two more sides that were obscured in the collapse (or explosion) of the main structure.

In looking closer at the South Massif, it became evident that the south side of the structure had collapsed inward or perhaps exploded outward (accounting for the hills behind it). If it was a collapse, it is possible that this forced the bright material visible on the north face out from under the structure, possibly through the dark depression (Nansen) near the center of the north face. It is highly unusual for a solid rock mountain (presumed by geologists to be a result of a volcanic uplift or ejecta from a massive impact) to collapse inward like this. Cinder cones on Earth frequently show evidence of some internal collapse, but these deformations are uniformly circular. The South Massif would therefore have to be one of the most unusual cinder cones ever discovered, if that's what it was. For one thing, cinder cones are exactly that, conical-shaped volcanic uplifts with distinctive rounded crater-like depressions ("vents") at their peaks. Rarely, if ever, do cinder cones take on geometrical shapes, especially hexagonal ones. Further, the "vent" in the South Massif—if that's what it is—is *square*. This is also highly unusual in any kind of cinder cone.

So if the South Massif isn't a cinder cone, then what is it? The official theory in the NASA geologic report on the area stated that the mountains may have formed in the immense impact that created the Sea of Serenity.[1] In simple terms, the mountains are huge chunks of rock that landed on the floor of the Taurus-Littrow valley and then were covered by "a thin volcanic ash unit." Of course, neither of these theories could possibly account for the astronauts eventually saw on the faces of the various massif's and mountains in the valley itself—but we'll get to that.

Close-up view of Apollo 17 landing site and hexagonal "South Massif." Dark spot at the base of South Massif near the center of top face is depression named "Nansen." White dot is Apollo 17 landing site. Note collapsed/exploded backside of the massif and implications of six-sided geometry. Image markup shows assumed hexagonal reconstruction (white lines).

The ostensible geologic reason for selecting this landing site was the opportunity to sample the "dark mantle" material that covered the valley floor. Supposedly, this would be from the earliest impact that formed the Sea of Serenity basin that the Taurus-Littrow highlands bordered. Lunar Orbiter images of the region also showed large boulders deposited along the bases of the mountains (particularly the South Massif). The bright patch of material overlaying the dark mantled valley floor extending from the north end of the South Massif all the way to the sinuous Lincoln Scarp was also of interest to NASA geologists. The bright material was presumably deposited from the event that led to the collapse of the mountainous structure. There were also several dark rimmed "halo" craters that were thought to be volcanic.

Using a landing approach that required him to drop the LM sharply down among the mountains at a steep angle and with little margin for error, Cernan guided the lunar module Challenger to a landing at the outer edge of the landing ellipse on December 11, 1972 at 19:54:57 UTC. The precise location was at 20° 11' 26.88" N x 30° 46' 18.05", or just beyond the magical 19.5° location.

Cernan managed to land the Challenger near a crater named "Camelot" and within driving range of the Nansen

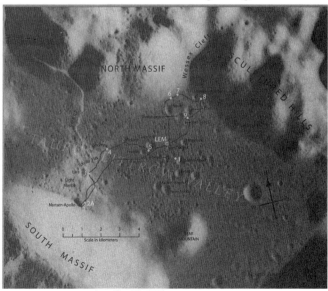

Apollo 17 EVA traverse map.

AS17-134-20391 with Bear Mountain, the East Massif and Mons Vitruvius in the foreground.

depression and the South Massif to the south and the North Massif and the "Sculptured Hills" to the north. Camelot was just one of several large and dangerous craters threatening a safe landing in the Taurus-Littrow Valley between the giant mountains and a more distant area to the east called "Mons Vitruvius." The crater "Isis," named by Mission Commander Cernan, was also was nearby just to add to the Egyptian symbolism.

The EVA (Extra Vehicular Activity) plans for the mission were the most extensive ever attempted for an Apollo mission. The science lobby inside NASA had pulled strings to get Harrison "Jack" Schmitt, a geologist, reassigned to Apollo17 after his previously assigned mission, Apollo 18, was abruptly canceled. Schmitt and Cernan had an aggressive schedule that called for them to unpack the Lunar Rover, emplace a number of seismometers and

High contrast enhancement of AS17-134-20391.

explosive charges at key points within the Taurus-Littrow Valley while traversing and exploring nearly 25 miles in total. They were also tasked to deploy a mysterious and classified experiment called "Chapel Bell," about which virtually nothing (to this day) has been revealed.

After unpacking and deploying the Rover and ALSEP instrument packages along with the American flag,

TV image of "condos" built into the side of Mons Vitruvius miles beyond the camera location.

Schmitt and Cernan headed to their first geology station, the nearby crater Steno. It was here that Schmitt and Cernan took the first in a series of panoramic photos that remain controversial to this day.

At first, AS17-134-20391 doesn't seem very interesting or controversial. A hand-held Hasselblad shot taken by astronaut Gene Cernan, it shows the view out of the moving Lunar Rover as the astronauts made their way to geology station 1. The mountains in the background are Bear Mountain and part of the East Massif. Again, this all appears fairly unremarkable until you zoom up close onto the base of the East Massif and see... condos.

In sharpening and enhancing the "mountain," you can plainly see rows and layers of evenly spaced, cell-like structures all along the front face of the East Massif. They appear to be built right into the face of the mountain, one row atop the other, exactly like

AS17-136-20767 and the Spar.

160

something that once resembled a step pyramid. These room sized cells are unmistakable signs of artificial construction, resembling numerous such structures in the American Southwest, similar to the abandoned Anasazi ruins.

Confirmation of this rectangular pattern was later found on high definition broadcasts of EVA-1 shown on NASA TV. Taken by the color TV cameras on board the Rover after it had come to a stop at geology station 1, they confirm the same anomalous geometric pattern can be seen on two different kinds of cameras and recording media.

It's also clear from the EVA transcripts that astronauts saw these bizarre, geometric patterns on all the hills and mountains surrounding the Taurus-Littrow valley. During one of the technical debriefings, Schmitt tried somewhat desperately to downplay them:

"It was a puzzle, seeing apparent lineations on the slopes of mountains." Schmitt said. "Some people, as I recall, did some simulations, building models, putting random roughness on the surface, and then dusting them and moving the light around, and they were able to create apparent lineations just with light position. Generally, I think, people don't feel that they represent any underlying structure, it's just an accident of dusting and

Close-up of the base of the spar showing it connected to an anchor point on the ground, miles from the Lunar Rover at the base of the East Massif.

Wide angle showing glass ruins over the East Massif.

lighting. The massifs do have layers in them, layers of debris, and I think the fact that you see what appear to be zones of blocks at the top is probably a layer of relatively hard material. But they really are gross layers."

Schmitt's comments are really nothing more than a variation of the old "trick of light and shadow" argument NASA has used for decades to explain away obviously unexplainable structures on various photographs of the Moon and Mars. But it's clear that Schmitt, a trained geologist, was trying to understand or explain how such features could be formed naturally rather than face the simpler explanation—they couldn't have formed naturally. They were artificial.

Further proof in support of this model (and the lunar dome model) came in the form of images scanned and downloaded from an Italian web site (since extinct) that displayed high resolution scans of photos taken on EVA-1 at around the same time and location. Apparently procured from European versions of early generation NASA negatives, these photos painted a completely different picture of the Taurus-Littrow valley than the one NASA has sought to advance

The upper tip of the Spar.

162

Contrast enhanced panorama.

for years.

Taken by Jack Schmitt a few minutes after Cernan's earlier photo of the East Massif and Mons Vitruvius, Schmitt's photo points in the same general direction as Cernan's and also shows the same geometric pattern on the face of the East Massif outcropping. Taken at geology station 1 near Steno crater, Schmitt's photo AS17-136-20767 also shows something else, a piece of semi-transparent spar-like glass structure which has sagged and collapsed over the mountain.

I must make it very clear here, this obviously structural spar does not appear currently on any NASA versions of this frame. It appears only on the Italian version apparently procured from a European scientific archive. Because of this, some people have sought to downplay it as a scratch on the print or negative, but this can be discounted by several observations. First, if you notice the spar follows the geometry of the light on the Moon that day; brighter and more reflective in the areas where there is ample sunlight and nearly invisible in the darker shadowed areas of the photo. It also follows the contours of the East Massif itself and seems to be anchored to the ground by a thin connecting wire. "Scratches" simply do not follow the contours of terrain or the geometry of the lighting in a photograph. If it was a scratch and not actually on the surface of the Moon, it would be the same brightness along its entire length.

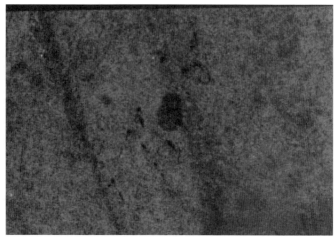

A hole punched in the glass structure over the Taurus-Littrow valley.

There is also the question of the photo itself, which is of much higher quality than many NASA scans available today. This is because most of the photos on official NASA sites are copies of copies of copies, many generations removed from the originals. These Italian images appear to be more akin to Ken Johnston's first generation prints in quality and detail. This makes the "scratch" explanation even more unlikely.

As part of the research for *Dark Mission*, we subsequently assembled a number of other scanned images from the European archives and found that they showed a far different version of the area around the East Massif than could be gleaned from current official NASA imagery. Under contrast enhancement, instead of the dark, pitch black sky that would be expected was instead highly organized, geometric blanket of light above the East Massif. This three- dimensional grid of semi-transparent structure looked exactly as would be expected After eons of bombardment from meteors and other debris. On several of the frames not only can very dark, spar-like vertical struts be seen, but there are even obvious areas where meteors have punched holes through the protective glass domes over the area.

Clearly, the "spar" itself is one of these vertical supports which has fallen from the matrix of glass overhead and collapsed down on the East Massif below. A close up of the upper part of the spar shows it still loosely connected to this fuzzy background material overhead. It's also clear from these other images that the overall structural matrix of the glass dome over Taurus-Littrow remains intact but clearly has seen better days.

Ultra close-ups of the three-dimensional matrix show that it has a wavy, layered quality to it that is entirely consistent with multi-layered glass meteor shield. There are also some fairly obvious points where huge impacts have punched holes in the glass medium.

Taken together, all of these anomalies and structures make a very strong case for the presence of an ancient alien base in the Taurus-Littrow valley. From the odd room-like structures built into the East Massif to the Spar sagging on top of it to the massive glass matrix suspended over it, it is clear NASA had high motivation to take the risks associated with landing there. But this still was only EVA-1. What the astronauts found and did the next day, on EVA-2, would be even more extraordinary.

EVA-2

The next day after their relatively short EVA to Steno crater, EVA-2 called for Cernan and Schmitt to head pretty directly for the South Massif and the odd depression named Nansen. After that, they were scheduled to visit locations on the Lincoln Scarp and then stop at Shorty Crater, one of the key targets for the mission.

In looking at the orbital images of the landing site, it becomes quite obvious why these locations (and EVA-3's planned visit to the "Sculptured Hills") were so important to NASA. Enhancements of the Apollo landing site orbital images show the same rectangular, geometric patterns on the faces of virtually all the hills and mountains in the landing vicinity. But the South Massif was something even more special.

Photos of the landing site taken by the high resolution panoramic camera on the Command Module from Apollo 15 show the entire South Massif is criss-crossed with highly unusual geometric patterns, both on the visible face of the massif and in the hollowed-out interior. In fact, looking at these photos of the

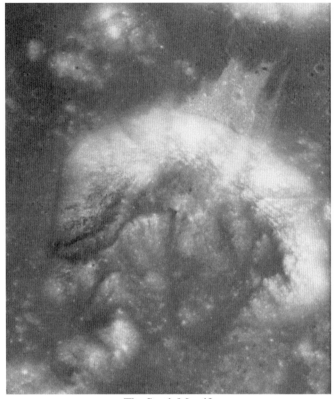

The South Massif.

supposed "vent hole" on the South Massif, the whole thing looks very curious, as if it did not collapse in on itself so much as it exploded outward, from the inside.

These close-up enhancements once again show the all too familiar cellular, geometric patterns all over the former interior of the Massif. It's as if we were looking at literally millions of small rooms which had been suddenly and dramatically exposed by the blast that blew the back half of the mountain off. In fact, that probably *exactly* what we're looking at.

Some geologists and other researchers (like Keith Laney) are of the opinion that the bright blanket of material around Nansen was created when debris from the blast that blew off the back half of the mountain pushed material out through the bottom of the South Massif, perhaps actually creating Nansen in the process.

As we described it in *Dark Mission*:

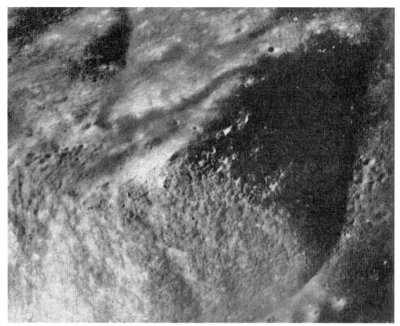

Room-sized cells on the South Massif.

Officially listed as a crater, Keith Laney has shown categorically that Nansen is nothing of the kind. Recon photos of the South Massif show Nansen as a V-shaped depression at the base of the massif, over which the "rim" of Nansen seems to be an overhanging shelf. Views of Nansen strongly imply that it is a hole in the base of the South Massif, possibly an entrance point (or exit wound) into the mountain. Certainly, if there were anything unusual about the South Massif, Cernan and Schmitt would be able to spot it either from "Geology Station 2," which was at the base of the South Massif atop Nansen, or on their way up (or back from) the station.

Everybody at NASA certainly seemed very excited about prospects for the day 2 journey to Nansen. Excerpts from the Apollo Lunar Surface Journal show that as the astronauts prepared the rover for the second EVA, Mission Control and Schmitt had this quick exchange:

141:02:06, Parker: "And, Jack …you might want to

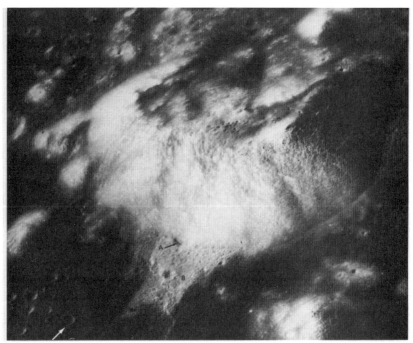

Apollo 15 pancam image of the South Massiff showing Nansen (A) and the bright debris blanket around it (B)

shoot off a few five-hundred-millimeter frames of the North and South Massifs, if they look interesting…"

141:02:27, Schmitt: (Incredulous) "If they look interesting!? If they look interesting!? Now, what kind of thing is that to say?"

After a quick stop to lay some seismic detectors down at different points along the valley floor, Schmitt and Cernan headed straight for the South Massif and Nansen itself. As they neared the Mountain, Cernan again noted the linear geometry on the surface of the mountain:

141:52:03, Cernan: "Jack, can you see over there to the left …of the South Massif where you get those impressed lineations? See them going from left upward to the right?"

141:52:11, Schmitt: "Yeah. I see what you mean; right."

141:52:14, Cernan: "That's what I saw out my window." (Here, Cernan is alluding to observations he'd made on the approach to landing the Challenger).

141:52:15, Schmitt: "Yeah, they go obliquely up the slope."

141:52:20, Cernan: "They're more like wrinkles, they're linear wrinkles."

141:52:22, Schmitt: "Yeah. Crenulations, you might say, in the slope that looks something like those I saw from orbit, looking in the shadowed area... or, at the edge of the shadows."

142:12:30, Cernan: "Jack, look at the wrinkles over there on the North Massif."

142:12:34, Schmitt: "Yeah. There's no question that there are apparent lineations all over these Massifs, in a variety of directions. Hey, look at how that Scarp [sic] goes up the side (of the North Massif) there. There's a distinct change in texture."

Later images, along with the orbital photography, have confirmed that the "lineations" are not tricks of light and shadow, but

Close-up of Nansen (A) and the bright debris blanket around it (B)
Nansen was the primary target of the Apollo 17 day two EVA.

"Lineations" on the South Massif.

undeniably real features of the mountains in this Taurus-Littrow Valley. Schmitt's inability to explain the "apparent lineations" stems from the fact that this kind of geologic layering is almost always associated with sedimentary deposition caused by standing water. Since no water has ever flowed or pooled on the Moon, such natural geological explanations are totally untenable. These types of lineations can also be caused under rare conditions by lava flows, but again the lineations on the massifs were literally thousands of feet above the ancient lava pool on the valley floor, making that explanation even less likely than water. Remember, NASA's geologic model was that these mountains were deposited on the Taurus-Littrow plane as ejecta from a massive impact event.

This leaves only one viable explanation for the regular,

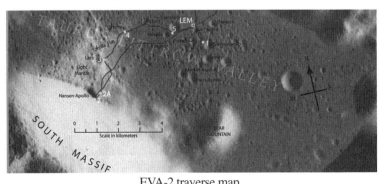

EVA-2 traverse map.

170

Composite view looking north of two frames (AS17-138-21058 and -21059) showing the view of Nansen from above, near where the Lunar Rover was parked. The ridge running through the center of the image is the shelf seen in orbital photos. No publicly released views exist looking directly into Nansen from the north.

repeating geometric patterns visible on the faces of all the "massifs" in the Taurus-Littrow valley—architecture. That this was difficult for the astronauts to come to grips with is unquestionable from the transcripts. But that observation is only the beginning of the mysteries that the South Massif held.

On the approach to Nansen, Schmitt, who has the task of taking a photograph every thirty yards or so during the traverse, suddenly stops taking pictures, saying at 141:56:24: "Holy cow! I'd better slow down my picture taking." Why exactly he would stop taking pictures of the primary target of that day's EVA is anyone's guess, but the two astronauts took literally hundreds more photos during the rest of the EVA. It was just then interior of Nansen they didn't seem to be interested in taking pictures of. This is in spite of the fact they had plenty of opportunity to do so:

142:42:21 Cernan: "Boy, you're looking right into Nansen."

A few moments later Cernan parks the rover at geology station 2, up on the shelf above Nansen and facing north, so the TV cameras cannot see directly into the opening at the base of the massif. Instead, we can only see down into Nansen from above and to the south of the curious opening. Schmitt and Cernan could

171

however have seen directly into Nansen once they dismounted the Rover and began moving around the area. It's pretty clear from the transcript that they were astounded by what they saw:

> 142:44:27 Schmitt: Look at Nansen! My goodness gracious.

In looking at the only available photos of Nansen, it is hard to figure out what was so exciting to Schmitt, unless he saw something off camera that was far more intriguing than the simple gray landscape we have been shown.

After dismounting, the astronauts activated the TV camera. Unfortunately, when we did get the television picture, it showed us little to nothing. Since they had parked the rover on the shelf above Nansen (the "entrance" to the South Massif), all we had was a view looking back to the north toward the light mantle avalanche runoff. After some minimal time spent on housekeeping tasks, the astronauts disappeared off camera for most of the next 20 minutes. In fact, they are out of sight of the camera for fully 85% of the entire sixty-four minute visit to the upper shelf of Nansen. This would have given them plenty of time to descend the hill and investigate the interior of Nansen, including examining the opening below the overhang.

As they are nearing the end of their time at the geology station, Cernan stops (off camera) to take some pans of the view from the base of the massif:

> 143:22:08, Cernan: "Well, I have some good pictures of Nansen, anyway, and... (long pause)... You know, I look out there, I'm not sure I really believe it all."

A bit later, completely out of context, Schmitt seems to address their "off-camera" time to mission control:

> 143:27:11, Schmitt: "We haven't had a chance to look around anymore than you've heard."

> 143:27:14, Parker: "Okay."

Is this an indication that they *didn't* descend into

Nansen during their off-camera time? And what was so unbelievable about Nansen that Cernan had to mention it? Certainly, there is nothing in the released images of the station that implies anything unusual.

Next, Cernan and Schmitt drove to station 2A, just a few hundred meters back down the slope they had ridden up to get to the lip of Nansen. From this perspective, they could have had a perfect vantage point looking directly into the hole leading into the South Massif.

Amazingly, there are officially no pictures taken of Nansen from this perspective, looking directly into the opening under the ledge and back up at the previous geology station. However, at station 2A, Cernan turned the rover in a circle, so Schmitt could take a panoramic view of the area and the valley below them. As he did so, there was this exchange:

143:50:20, Cernan: "Wait a minute. Wait a minute. Okay. Let's take one from right here. I want the whole thing. (Pause.) You ready to start?"

143:50:26, Schmitt: "Yeah, I got it."

143:50:27, Cernan: "Start taking. Take the whole thing."

143:50:54, Cernan: "Isn't that something? Man, you talk about a mysterious looking place."

143:51:03, Schmitt: "They can cut some frames—some parts of those pictures out—and make a nice photograph."

According to the transcripts, the pan "turned out to be relatively uninteresting because of the sun glare." Published images do not seem to be a full 360° pan, because if they were, the entrance down into Nansen would be visible. In fact, it's nowhere on the published sequences of this area. And how does Cernan's statement "Man, you talk about a mysterious looking place" jibe with the "relatively uninteresting" description given in the transcripts? This whole area seems to be fascinatingly anomalous, so much so that Cernan stopped to make sure Schmitt got a complete photographic record of it. A record, incidentally, which now does not correspond to the descriptions of the astronauts at that time. And what about Schmitt's nervous talk about cutting "some parts of those pictures out?" What does he see that NASA doesn't want the folks back home to see?

Maybe this. At some point after they arrived at geology station 2 above Nansen, Schmitt took a series of photographs that show the floor of the Lunar Rover vehicle, frames AS17-135-20676 to 20679. But the last photo in this magazine, AS17-135-20680, is listed as "LRV Floor? Sunstruck" in the NASA Apollo 17 image library.[3] It is a very noisy and over exposed image. But when the noise is removed, it takes the shape of a very pyramidal looking object, which may or may not be part of the Rover itself.

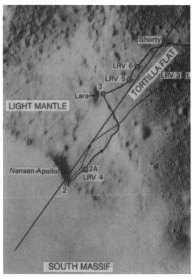

EVA-2 traverse map showing the locations of geology stations 2 and 2A at the base of the South Massif. Note that station 2A would allow a direct view into the Nansen opening.

Is it possible that Schmitt was supposed to take the pictures looking directly into Nansen from the north near geology station 2a, and forgot to change film magazines? If he had been tasked with taking the photos of Nansen from the non-public perspective, the one that we were never supposed to see, maybe there was a special "black" film magazine he was supposed to use? Maybe the "Chapel Bell" film magazine?

If this is true and Schmitt simply forgot about the one errant frame he took of Nansen's interior, then it is possible that this is one of the long sought images that were part of the classified part of the mission to Taurus-Littrow. Certainly, the provenance of this photo implies at least some desire on NASA's part to discourage anyone from ordering it or looking at it. While it is listed in the Apollo 17 image library, it is misidentified as being part of magazine 135 (G). In fact, it might be part of magazine 136 (H). So the correct frame number could be AS17-136-20680. However, you will not find it on any NASA web site with this designation, and the Apollo17 photo index from the

1970's lists that frame number as "blank," which it obviously is not.[2] And, on the NASA Lunar and Planetary Institute web site, the frame doesn't exist at all. Frames AS17-135-20680 (magazine G) and AS17-136-20681 (magazine H) are simply missing.

As you read back through the transcripts, it's clear from the

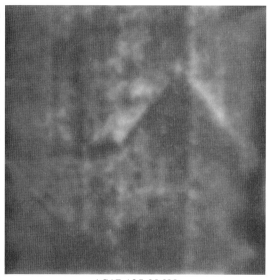

AS17-135-20680

comments of the astronauts that something is amiss. First, Cernan is concerned that they can see directly into Nansen, so much so that he parks the rover up over the lip of Nansen so the TV cameras can't see directly into the opening. Schmitt—the trained scientist—is so astonished as he looks into Nansen for the first time that he exclaims, "Look at Nansen!" Cernan goes on to describe the whole area as a "mysterious place," and "unbelievable," and Schmitt cryptically informs Mission Control that the astronauts "haven't had a chance to look around anymore than you've heard," despite their thirty-plus minutes off-camera.

Later, as they take pictures that should show what lies beneath the shadowed "overhang" of Nansen (which is clearly visible from orbit), they joke about cutting out certain parts of the pictures. And in all this time, not once did they take a picture showing Nansen from a vantage point that would reveal the interior of the "crater." They evidently just missed the opportunity, and nobody at Mission Control decided to ask for such a picture.

The reality, however, is that between their off-camera time and the traverse to Station 3, the astronauts would have had plenty of time to descend from the rise and examine and photograph the interior of Nansen. That these photos could have been simply lifted from the photographic catalogs is easy to conceive, especially given the shenanigans around AS17-136-20681. It's possible that Schmitt and Cernan found that getting into the South Massif through Nansen was impractical. There were descriptions of a great deal of debris in the crater. Or perhaps they tried and failed, leading to Schmitt's admission that they didn't get to look around as much as they wanted to.

Whatever the case, they were on a tight timetable to get on with the other stations. What they could not have known, however, was that they were also on a collision course with an even more unbelievable and mysterious encounter, with "Data's Head."

Data's Head

"Mr. Data, your head is not an artifact."

– Commander Riker, from the *Star Trek: The Next Generation* episode "Time's Arrow."

I still distinctly remember the first time Richard C. Hoagland showed me the picture of "Data's Head." I don't think I will ever forget it. I was sitting on the couch of his office in Albuquerque, New Mexico, looking at some images he was showing on a big blank wall with a projector. This was the last of several "Dark Mission" summits we'd had to settle on the contents of that book in 2006, and I was anxious to get back to my home in Los Angeles, not the least of which was because L.A. has two things Albuquerque does not; moisture and oxygen. I was tired, cranky, and ready to get to bed. "Don't be so restless," he said mysteriously. "I've got one more thing to show you."

Yeah.

There, slowly but surely, building the tension in the way that only he can, he took me through some of the data that you have just seen and read. Then we began to focus in on the last major stop of EVA-2, Shorty crater.

Along the way to Shorty Crater, a black, halo-rimmed volcanic crater on the outskirts of the light mantle material from the South Massif avalanche, Schmitt and Cernan stopped to take some samples along the rim of another small crater, which came to be known as

"Ballet Crater." It was thus dubbed because Schmitt had taken an exceptionally entertaining spill there while trying to retrieve some sample bags. The stop, which was supposed to last twenty minutes, was made to take a double core sample, get a gravimeter reading, and take some photographic pans of the general area (again, notice that Nansen seems to be the only stop on EVA-2 that what not photographed in extensive detail, at least officially). It took nearly thirty-seven minutes for the astronauts to complete their tasks at Ballet Crater, and from there it was straight on to Shorty, which was a primary stop for the EVA along with Nansen. Upon arrival at Shorty, the astronauts took care of some housekeeping chores, and then got their first look at the crater itself. In looking at the transcript of EVA-2, what they saw apparently astounded them:

> 145:22:22, Schmitt: "Shorty is a crater, the size of which you know. It's obviously darker rimmed, although the fragment population for most of the blanket does not seem too different than the light mantle. But inside... Whoo, whoo, whoo!"

Schmitt's description seemed to imply that while Shorty was relatively unspectacular on the outside, the area inside the crater was, at the least, very interesting. Unfortunately, when the camera started up, it was pointed at the distant South Massif. It stayed positioned there as Schmitt moved away to take a panorama of the crater. Later, Cernan took a panorama of the entirety of Shorty crater (NASA frames AS17-137-20991 to 21027).

My first mind-blowing moment was when Richard showed me his color enhancement of Cernan's panorama. It was filled with stunning colors; blues, purples, pinks and greens, yet the white of the space-suited astronauts were pure white. It was then that he told me of his theory of light bending prismatic effects of the glass-like ruins overhead. I was even more stunned when I saw the famous orange soil that Cernan had spotted.

The orange soil turned out to be highly oxidized titanium, but it was nothing compared to what he showed me next inside the crater

Black and white version of NASA frame AS17-137-20990, showing the heavily oxidized orange soil near Shorty crater.

itself. I could instantly see what had made Schmitt so excited when he first looked down into Shorty. Rather than the normal looking rocks, boulders and debris I had been expecting, my engineers eye immediately focused on objects inside the rim of the 100-yard-wide crater that were quite obviously mechanical in origin.

Mechanical objects have certain features and tell-tale design aspects that clearly differentiate them from rocks and other random forms of nature. These include flanges, opposable handles, vents, heat sinks and all manner of connecting hoses, tubes and fixtures. These types of objects, these mere "rocks," as NASA would claim, were scattered all over the inside of Shorty crater. Most of them had clear (to my eye at least) mechanical features, fixtures and functions. We have a guiding principal in engineering design, which we call the "3 Fs," Form, Fit, and

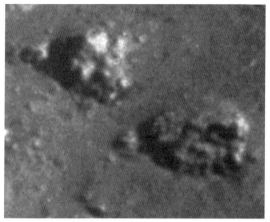

Mechanical debris in Shorty crater.

Function. If whatever you are adding to a part or an assembly doesn't enhance one of those characteristics, it doesn't serve a real purpose other than to make you feel clever. The pieces of machinery I saw in Shorty crater all had discernible form, fit and function, and that is what made it clear to me I wasn't just looking at rocks.

Then my eye went to a large, box like object which seemed to have welded venting tubes all around the outside of it. I thought it looked a bit like a big mechanical turkey or even the torso of some formerly functioning device or junction box. I could even see a clevis or fit point at the front where something else might have been attached by a pin.

Richard noted all this, and then said "Look up."

That's when I saw it, just above the turkey-torso like object. Something that absolutely should not have been there. A head. A disembodied head in a crater. On the Moon. Data's head.

More mechanical debris.

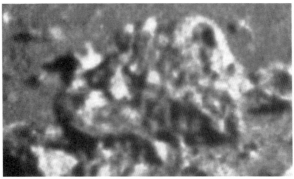

The "Turkey."

To say *my* head exploded when I saw this object is to do the opening scene in the movie "Scanners" an injustice. I literally could not believe it. But after looking at photo after photo of it, each more intriguing and revealing than the next, I could no longer deny the reality of it, especially considering the extensive mechanical context it was found in. There was somebody's head on the surface of the Moon. Or at least, *something's* head on the surface of the Moon...

After a brief discussion, we were in quick agreement that neither this object nor anything else in Shorty crater could be of biological origin. The airless vacuum and harsh radiation environment would have long before broken down any kind of living tissue into dust. No, to have survived this long on the exposed lunar surface, this and everything else we were seeing had to be mechanical in origin. So it had to be something mechanical—a robot's head. That mind blowing conclusion was only confirmed by color enhanced close-ups of what we were now calling "Data's Head," in honor of the android character from the dismal *Star Trek–The Next Generation* TV series.

You see, under color enhancement, it's easy to see that Data's Head has a bright red stripe painted on it. Obviously, that's a feature most "rocks" don't come with. The red stripe is plainly visible even without enhancement on several photos Schmitt took of the interior of Shorty crater, in spite of the claims by some critics that it isn't really there, or can't be reproduced.

As I looked at this undeniable mechanical artifact on the surface of the Moon, I kept thinking back to the episode of *Star Trek–The Next Generation* where the crew of the Enterprise finds the android

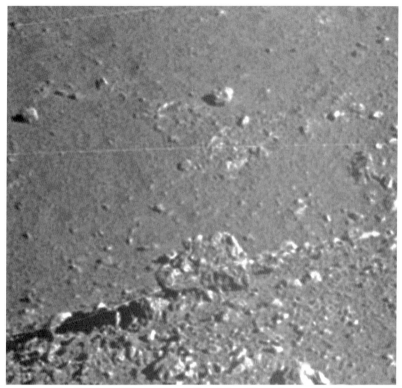

The interior of Shorty crater. Data's head is at the top middle.

lieutenant Data's severed head in excavated ruins in San Francisco. Was this object somehow symbolically connected? Was this some artifact left by the Ancient Alien civilization that once flourished on Earth's Moon? And if it was, did Schmitt and Cernan see it? And more importantly, *did they bring it back*? They certainly could have, as it is exactly the same size as a human head and would have fit into the

sample bags fairly easily. The question of whether they had the off camera time to retrieve it is something else entirely.

To be sure, they continued to have opportunities to retrieve ancient alien technology from the surface of the Moon even after Shorty crater and Data's Head. There are photos from

Composite close-up of Data's Head.

181

NASA frame AS17-140-21409.

the third day and EVA-3 which also show what appear to be pumps, engines or some other kinds of mechanical debris scattered all over the Taurus-Littrow valley. Did they collect some of them and bring them back?

In the end, all we know for certain about the Apollo 17 mission is that it was unusual in almost every way, from its landing site to the lack of photos of the interior of Nansen to the mysterious and still classified "Chapel Bell" experiment. But from the vantage point of 40 years in the past, it certainly appears that this mission was designed to investigate and retrieve technology left over from the Ancient Alien habitation of the Moon in the distant past. The question is, to what end?

[1] http://history.nasa.gov/alsj/a17/a17pp-geosynth.pdf

[2] http://www.hq.nasa.gov/alsj/a17/images17.html#20942

[3] http://www.hq.nasa.gov/alsj/a17/a17.photidx.pdf

CHAPTER 8

CLEMENTINE, TYCHO AND THE FACTORY

Before we wrap up, I feel it's important to look at some other research and some of the newer missions that have more recently returned to the Moon.

In 1994, the Naval Research Laboratory, the Ballistic Missile Defense Organization and NASA jointly sent a probe named Clementine to the Moon and took thousands of new images of our only satellite for the first time in decades. We'll look at some of those images in a bit, but what was most provocative about Clementine was that "she" carried an excellent, high resolution digital imager which was state of the art for the time and could have afforded views in much greater definition than any of the 1960s and 70s Apollo era missions did. Unfortunately, in a move that brings up all of the "military intelligence" jokes I can think of, they didn't get any hi-res images to speak of. At least, none that they've released to the public. The supposed reason? They left the lens cap on.

I'm not kidding.

The ostensible reason for this oversight was that the NRL had scheduled Clementine to rendezvous with an asteroid named 1620 Geographos, and they wanted to "save" the hi-res imager for that mission. As a result, most of the hi-res images available to the public which funded the mission are pitch black. However, there were some interesting medium resolution images taken, and a couple of them got me started on

Clementine image LBA5904z.193. Bright star-like object at top is the planet Venus.

lunar anomaly hunting way back in the day.

The first image was actually taken with Clementine's low-resolution "star tracker" camera, which was used to navigate and position the spacecraft in its orbit around the Moon. On its 193rd orbit (April 1st, 1994, to be exact), Clementine was instructed to take an image of the lunar disk as the sun was receding behind it, in order to capture an image of the solar corona. It did that quite adequately. But it also captured something

Close-up of Clementine TLP.

else; a stunning Transient Lunar Phenomena.

There, right on the edge of the lunar disk and exactly where you'd expect to see a bright specular reflection (if the transparent lunar dome model is correct) was a bona-fide, in your face TLP. Never again would anyone have to debate the reality of the phenomenon. A second image,

LBA5905z.193, was taken one second later and confirmed the bright, glowing spec was not just an accident of the imager and that the flare of light lasted for a measureable period of time. Close-ups showed this strange reflection was right on the limb of the terminator, exactly where it would have to be if the glass structure model of the Moon was correct.

So if nothing else, Clementine had shown us that TLP's do exist, are still happening and while they remain officially mysterious, this one's appearance was exactly consistent with the glass lunar dome theory we've discussed in this book.

The truth is, I really didn't expect (or get) much from Clementine. Its mission and instruments were classified, and after the shenanigans with the hi-res imager, I was not expecting much in terms of the data that would be offered to the public. But then one day in browsing some press release images of the crater Tycho, I got quite a surprise. Tycho has always intrigued me because it such a prominent feature on the lunar disk and it was the location of the Monolith in Stanley Kubrick's *2001: A Space Odyssey*. So, given its storied fictional history, I decided to have a look at what was in the bottom of the crater. What I found not only freaked me out, it led to me eventually writing this book.

The image in question was a mosaic showing the central peak and north rim of the crater in three different wavelengths. The first two wavelengths showed a polygonal fracturing pattern in the floor of the crater, which is pretty normal for an assumed impact crater, and also reflected the different material compositions of rocks in the crater floor. But the third image, taken with Clementine's

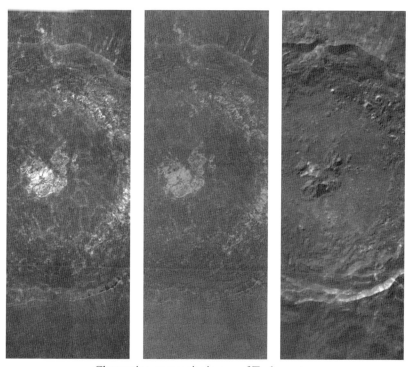

Clementine composite image of Tycho crater.

185

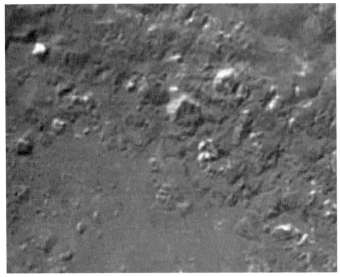

The Tycho Village.

Red/Green/Blue (visual) filter showed something quite different. There, mounted on the northeast rim of the crater and overlooking the central peak was some really weird looking stuff.

My eye was caught by them because they had completely reflective properties than the surrounding terrain, standing out as bright white reflectivity in the false-color image produced by NASA. But it was also their shapes. They looked a whole lot like a

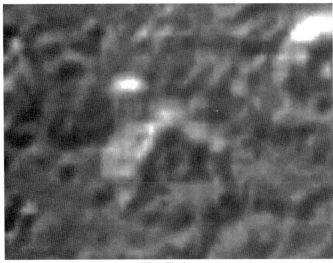

The Chalet.

186

pyramid, a couple of buildings and some heavy equipment.

For comparison purposes, I labeled these objects the "Pyramid," the "Chalet," the "Geo-dome," the "Longhorn" and the "Backhoe," and I called the whole area the "Village." This does not mean that I am suggesting that these are in fact a backhoe, chalets, or a true pyramid. Remember, Tycho is over 65 miles across, so whatever these objects are they're immense. In fact they are far too big to likely be what I have labeled them as. However, all these objects stand out based on their brightness relative to the surrounding terrain, their unusual geometric shapes, and their proximity to each other.

The "Chalet"

This object I dubbed the "Chalet," looks somewhat like an A-frame building. Note that it appears to form fit on the left hand side matching the terraced terrain. The roof is also far brighter than the surrounding ground, indicating it is made of a different material. The roof also seems to overhang the flat faced supporting structure, and dark, regularly spaced "windows" are visible in a pattern on the front face. The "chimney" has several dark vertical striations, hinting at some sort of venting system. They are almost certainly not image artifacts, as they end where the chimney meets the roof. The chimney also appears to have a domed tip with a flange.

The "Geo-Dome"

The accompanying object, the "Geo-dome" also displays comparably peculiar characteristics. Roughly hexagonal in shape, it too seems to jut from the hillside as if it were the tip of a much larger sub surface object or was actually "dug in" to the side of the hill. It as well has an apparent overhanging roof with an exposed, flat-faced front "wall" facing out to the crater. It

The "Geo-dome."

187

also has "windows" similar to the Chalet and a possible entrance at the base. It is set apart from the background by the brightness of the roof and its geodesic shape, similar to late 1970s concepts of solar homes.

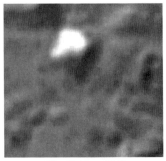

The "Pyramid."

The "Pyramid"

Slightly northwest of the "chalet" is this enigmatic faceted object. Again it is very bright and appears to have a 4-sided pyramidal structure. The direction of the shadows makes determining the details of the underside impossible, but the object does appear to be generally 4-sided and symmetrical.

The "Backhoe"

This object resembles a tractor or bulldozer with a drooping scoop set off to the right. It seems to have an opening in-between the "arm" and the base of the main body. There are cavities beneath the object in shadow, separated by a "post" between them.

The "arm" seems to be made of as many as 8 individual components. The shadow indicates this is a solid vertical structure and its brightness against the background implies it is not made of the same material as the surrounding terrain.

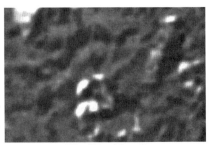

The "Backhoe."

The "Longhorn"

This looks to be a basically symmetrical object with 2 central "nodes" and curved arms extending from the central body. There appears to be some underlying support just to the left of the right hand curved "arm", but the central spherical "node" looks to be above the ground, judging by the shadow beneath it.

Note that is also sits in what appears to be an excavated "pen"

188

or platform. I am not aware of any accepted process that could account for this object forming naturally. In an effort to understand the shape I was seeing, I rendered the object and the pen in a 3D CAD system as I saw its various pieces and components.

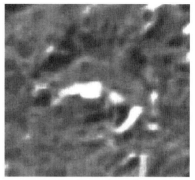
The "Longhorn."

Each of these objects on their own could potentially be explained away by some exotic natural process. However, due to their grouping in such a small area and varying visual characteristics, it is unlikely one explanation could encompass them all. These objects all have the look of machinery or constructs, as opposed to the simple boulders and cracks which should dominate this landscape. Indeed, the very geometric complexity of these objects argues for their non-natural origin.

If we had been stuck with just the Clementine data from 18 years ago, we might never have known any more about these weird objects on Tycho's rim than we do today. However, in 2009 NASA sent the Lunar Reconnaissance Orbiter to the Moon, with a hi-res camera better than anything that had been sent there before. They very quickly made the floor of the crater Tycho a target of this wonderful instrument, but early views showed the northeast rim in shadow and later images were of no better quality than the Clementine data. I assume much better versions of the LRO Tycho images do exist; I just haven't been able to find them as of this writing.

As to the ones that have been publically released, well everything still seems to be there, and my earlier findings are certainly not refuted by what they show. They still stand out because of their brightness and unique visual characteristics. There still

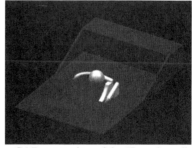
CAD recreation of the "Longhorn."

189

The Tycho Village from Lunar Reconnaissance Orbiter.

appear to be entrances into some of the objects, and the backhoe and longhorn are still just plain weird. And then there's that pyramid looking thing...

I've always found it very curious that these objects on the northeast rim of Tycho disappeared in the spectral analysis done during the Clementine period. Perhaps only a landing—manned or otherwise—will settle their origins once and for all.

Hortensius

Another region of interest on the Moon that I was hoping to update was the area around the craters Hortensius and Hortensius C. Back in the late 1990s, I worked with Steve Troy on looking over the area for my old Lunar Anomalies web page. Steve had found some very interesting geometry in the area between the two craters, and had forwarded the excellent Lunar Orbiter medium resolution image to me for study.

Hortensius is an equatorial [6 degrees N, 28 degrees W] nearside, 14 km-wide crater just southwest of Copernicus. The region

NASA frame LO-III-123M. Prominent craters in the middle of frame are Hortensius and Hortensius C.

is dominated by a series of volcanic domes stretching several hundred miles north of the crater itself. When I originally got LO-III-123M and the sectionals from Steve I wasn't all that excited. Sure, there were some interesting patterns and the "volcanic" domes of this area don't look all that volcanic, but I had basically decided this would be a backburner project. Then I looked at the sectionals.

They revealed a vast expanse of regular geometric patterning that seemed to flow in the same directions, parallel to each

191

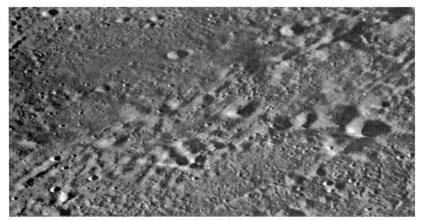

Wide angle view of complex, geometric patterning between craters Hortensius and Hortensius C.

other all the way across the frames and crossing at very regular perpendicular angles. They were also not in line with the grain of the film and were substantial enough to be blatantly obvious with a magnifying glass. You can even note them on the highly compressed crude scan above. Volcanic fracturing can certainly cause parallel patterns, but perpendicularity over a vast (100 square kilometers) area is far harder to explain away as volcanic. In addition, the patterning seemed to be more like channels or tubes rather than a fracture pattern.

The really weird thing though was that a lot of this pattering was *on top* of the feeble ejecta blanket around Hortensius (the big crater in the lower middle) itself, hinting that it either came *after* the formation of the crater or was only partially obliterated by the thin ejecta layer.

Closer view of the "Factory." Note triangular bunkers and "access roads."

Now, completely setting aside for the moment the strange lack of "spew" from such a large impact, my eyes were drawn to the area to the right of Hortensius between it and Hortensius C, the smaller (about 7 km diameter) crater at the image's right edge. In here, I noted (as Steve had) some very significant geometry—and some overt structures.

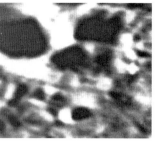

The bunkers in high contrast.

The "Factory Complex"

This stunning region is about 4 x 3 kilometers, judging by the size of Hortensius C. It is dominated by triangular "hanger doors" leading to semi-recessed bunker like structures, and a stunning black box shaped object the size of a 10 story building. In addition, the pronounced right angle

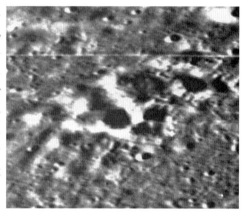

High-contrast view of the Factory.

pattern in the foreground is reminiscent of barely covered tunnel network. There is also a striking symmetry in the exposed areas (the T-shaped feature behind the building for instance).

What struck me first were rows and rows of archways and entrances that were at different levels of the structure. There seemed to be access roads, entrances and hangar doors all along this area. Contrast enhanced close-ups showed some even stranger structures, pipes and tube networks all around the area.

This series of raised, parallel triangular openings are recessed into the surrounding terrain, and compare favorably to the camouflaged bunkers here on Earth. This reinforces the impression of a factory or storage complex. Note also the straight line behind the first set of

Black object (center) is the "Lincoln Memorial."

"bunkers." Organized facilities display regular, repeating patterns and identical features over wide distances. Natural formations are far more random.

The shadows cast by the bunkers are inconsistent with oval cratering caused by ejecta impacts. To say this arrangement of objects is anomalous is a wild understatement. They are flatly inexplicable in a currently accepted or theorized geologic model.

There are parallel terraces all along the right side of this Factory and perpendicular striations (access roads?) around the bunkers. Conceivably, the underlying tunnel network could be lava tubes formed in the ancient past similar to riles and ridges seen in other regions of the Moon. However, it should also be observed that such tubes have been proposed as ideal locations for eventual human bases on the lunar surface because they provide easily sealed off cavities with natural protection from the harsh radiation and temperature variations that would be encountered. This would make modified "natural" lava tubes an ideal place to hollow out a sub-surface base of some kind.

The Lincoln Memorial

Just above and behind "bunker row" is a large, dark object

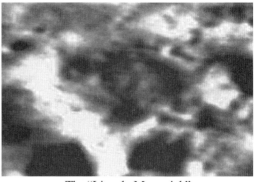

The "Lincoln Memorial."

194

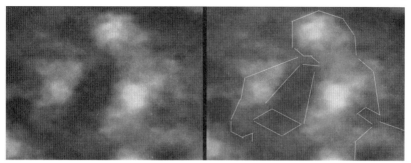

Marked up details of The "Lincoln Memorial."

that appears to be sitting exposed on a plank of sorts above a dark opening. It has a large, rounded object on top of it that is casting a shadow on the main body below. The right part of the plank appears to be connected to the central complex by a dark, cylindrical object.

Quite simply, there is no conventional natural explanation for this set of objects. Indeed, there is no conceivable geologic process or even set of processes which can account for these structures. It certainly isn't any kind of crater.

Notice that there is no discernible rim, and compare it with countless other examples of sharp edged impact regions across the lunar landscape. The "Lincoln Memorial" itself is a roughly cubic shape, with spherical nodes or buttresses at the base along the "plank". Lincoln also seems to have a "head" that is approximately spherical and casting a vast shadow over his "chest." To the right, a canister like object seems to be linking the "plank" to the main body of the facility. The "Truss" has cylindrical central body spanning the "plank" and the edge of the "terraces." It is anchored at both ends by a half-slot shaped end cap which extends to the ground on each side. The archway underneath the cylindrical shape is plainly visible. This object would appear to be a support or reinforcing member holding the "plank" up.

Note the shadow cast beneath Lincoln seems to be a hollow area with no visible support for the

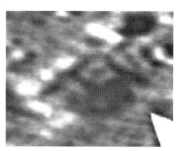

The Overhang.

195

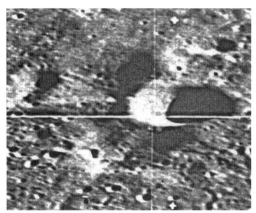

Crater with support struts and girders around it.

structure. The plank itself would seem to be only thing holding the Memorial up.

Ultra close-ups not only show the shadow cast by the head, but also what may be a rectangular platform beneath. This platform would be large enough to serve as a landing pad for a vehicle like the Lunar Module.

And there are other strange, architectural looking objects all over LO-III-123M. On the right side of the Factory complex is a dark opening with a disk-shaped object hanging over it.

Notice it is supported from the upper rear by a strut, and that the surrounding terrain seems to flow into the darkness below it, as if this were the entrance to an underground bunker or complex. Given its location adjacent to the "Factory", this may indeed be precisely that. Note the generally square shape of the opening, and again the even symmetry of the both the "Overhang" itself and chasm it seems to guard. There is also a light, dome shaped node just beneath the "Overhang" in the darkened area which may be indicative of light (from the nearly directly overhead sun) creeping through thinned areas of the "Overhang" itself, or possibly of some form of self-luminescence. A final possibility is that the "Overhang" has a light source in its central disk and this is projecting downward. There certainly seems to be a rounded "something" in the dark area below the opening.

The last area I'll cover from LO-III-123M is the area around the crater just east of (and adjacent to) the Factory. When I looked at the scans of the photographic negatives (which are much higher resolution than the current "high-res" scans on NASA websites) I was frankly, pretty astonished at what I saw. There, in a crater next to the Factory, was a bunch of structural debris.

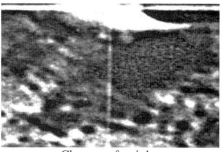
Close-up of a girder.

The crater itself is approximately two KM across and seems to be of the collapse variety. On closer inspection the area around it seems to consist of intersecting structural members barely poking through the regolith. They are very linear and have regularly spaced "lightening holes", or dark areas, in them.

As I studied the image I was struck by what appeared to be an edge to the crater on lower left side just below the framelet line. Note that there is a distinct rim in this portion of the image. Given the shadow length, I was also a bit disturbed by the dark zone just below and to the right of the crater rim. If this were a normal crater, the area should be illuminated rather than pitch black. It occurred to me that there must have been a collapse of the surface skin in this dark zone.

By all rights, this should be a lopsided partial crater with a significant buildup of collapse material in the dark zone. The question then becomes "What is providing the structural stiffness of the crater itself?"

The simple fact is that unless the crater is an extremely stiff structural object, (like I don't know, a satellite dish?) it should probably have collapsed into the recessed dark area. Given that it did not, I considered the winding, striated feature in the left center of the object as a possible support strut. It projects from the surface at a 45 degree angle and an apparently anchors under the rim of the crater.

Note also the parallel lines making up segments of the "Strut" are aligned at this same angle, rather than running along the direction of the surrounding terrain. It has all the earmarks of a spring loaded strut, similar to the earthquake absorption systems in modern office buildings. There is also an odd jagged edge on the right portion of the deeply shadowed area, as if a plate or board was broken by an impact. Underneath the "Strut" is an object easily recognizable as an I-beam or girder with lightening holes. This is similar to other such girders just north of the crater. It seems to be slightly stressed and bent under the "Strut." The truth is, this object has all the engineering characteristics of an I-beam. It has a central web, raised sides (or flanges) and equally spaced lightening holes. I believe that the middle lightening hole is slightly distorted by the vertical line (which may be a defect) passing just through the middle of it in the photo.

When it comes to the "ground truth" we talked about in chapter 4, the Factory area is, in my opinion, one the best *proofs* yet of ancient alien inhabitation of the Moon. Not only are all the features plainly visible in both the analog and digital data sets, they defy any reasonable prosaic geological explanation. The terraced, multi-level topography and recessed bunker formations are indicative of military installations in remote areas of Earth, and the "Lincoln Memorial" (which is the size of a ten story building) evokes comparisons with power plants or other large-scale facilities.

The other objects discussed in areas adjacent to the Factory are also well outside traditional Lunar geologic modeling. They appear to be part of a massive and well organized underground facility in the region, and they reveal damaged and exposed support structures that make architectural sense while simultaneously defying natural explanation.

But the fact is, there are still those who will refuse to believe that what they've been shown in this book is proof of the Ancient Aliens on the Moon hypothesis, or that NASA would want to cover any of this up. That why I've saved the best for last.

CHAPTER 9

THE ZIGGURAT, COPERNICUS AND BEYOND THE INFINITE

Ziggurats are massive Mesopotamian structures that were built thousands of years ago in what is today Iraq and modern Iran. Primarily used as temples and fortresses, ziggurats are stepped pyramidal structures architecturally similar to the step pyramids of Central and South America. They always had a specific set of architectural features that distinguished them from other pyramidal structures; a ramp or set of ramps leading to the top, high walls and a single ramp that led to a shrine or temple at the top, presumably to be used for rituals to the Gods that created or inspired them.

And there's one on the Moon.

Late in the process of writing this book, I came across this close-up view of what can only be described as a full, functional and only partially buried ziggurat on the Moon. I found it on an internet message board while searching for something else.[1] At any rate, the photo of the Ziggurat was listed as coming from Apollo 11 photographic frame AS11-38-5564. It is truly rare that a message board posting of this type actually comes with a frame

The "Ziggurat" from Apollo frame AS11-38-5564.

number, but this time I was in luck. The image of the Ziggurat comes from an orbital view of the lunar far side (remember, not the dark side), somewhere west of the crater Daedalus at 7.5° S / 175° E. This is interesting for a number of reasons, not least of which is that Daedalus is very close to 180° around the circumference of the Moon from the triangular crater Ukert on the front side. That means that the Ziggurat lies at what is called the "antipodal" point to Ukert, the point exactly opposite (or nearly so) on the lunar sphere.

At first blush, the image is impressive, showing a clear, pyramidal, walled structure with a central core (or temple, if you like) rising from the interior structure. The central temple has what appears to be a bridge or walkway leading into it, and the square temple housing has a dome on top of it. To the front are twin, angled ramps leading to the top of the wall. It bears a striking resemblance to the ziggurat at Ur, in southern Iraq, having all the same features if a somewhat different architectural arrangement.

When I did the side-by-side comparison, I was even more impressed with the resemblance.

Other comparisons with the artistic reconstructions of the ziggurat at Ur were even more compelling since the top few

The Ur Ziggurat.

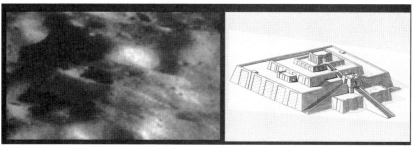

Daedalus Ziggurat (left) and artistic reconstruction of Ur Ziggurat (right).

levels of the surviving relic are now nothing but dust.

To me, this almost became the Holy Grail for the Ancient Aliens on the Moon theory. If this structure was real, there could be no doubt whatsoever as to its artificial origin. All I had to do was find it on AS11-38-5564 and confirm it with my own enhancements. After some pursuing of the internet, I found that this pyramid was fairly popular and that it had been spread around quite a bit. The original file that I downloaded was named "as1120pyramid20smallue2. jpg." I did not find any specific reference as to who initially spotted it or put it on the Net, although John Lear's name was mentioned in one message board. Lear however, denied that in a Facebook posting in July, 2012.

So after some searching, I downloaded the print resolution jpeg from the Lunar and Planetary Institute web site and began looking for the Ziggurat.

I didn't find it.

As it turned out, there was a very good reason I didn't find it. It wasn't there. At least, not in the form that it was shown on the web sites I'd downloaded it from. It took me quite a while, but I eventually did find the area in frame AS11-38-5564 where the Ziggurat *should* have been — it just wasn't there.

What was there was a fairly normal, bland, mundane looking lunar landscape.

So either the Ziggurat itself was a fake, or the current publically available NASA version of AS11-38-5564 was a fake that had been altered to cover up the presence of the Ziggurat. What I could see clearly on both images was the angled "ramps" that led up to the top of either the simple hill, or the structure, depending on which

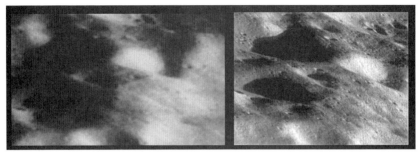

The Daedalus Ziggurat from as1120pyramid20smallue2.jpg (L) and from the current NASA version of AS11-38-5564 (R).

one was real. At least the ramps were real. Also, if the initial image of the pyramid was a fake, it was a very good one, with subtle details and the correct lighting to be a legitimate photo.

The question is, was it? I tried to search the Internet Archive to see if the NASA web pages had been altered over the years, and sadly *not* to my surprise I found that LPI/NASA doesn't allow their pages to be archived in this manner. In fact, I am probably

Full version of AS11-38-5564 (NASA) white square indicates location of the Ziggurat.

responsible for this myself to a certain degree. Back in 2002 during a major controversy over infrared images of the Cydonia region of Mars, I had caught NASA's Dr. Phillip Christensen in bald-faced lie about the images using the Archive. Since then, most NASA sites have opted out of the option of being periodically archived. Go figure.

At this point, I had no real way I prove the genesis of the Ziggurat image. It had no traceable lineage (unlike the Ken Johnston photographs) and I had never seen a full size version of AS11-38-5564 with the Ziggurat on it in order to do my own context work and enhancements. The available NASA digital images were possibly sanitized and there was no way to trace if they had ever been altered after being placed on the web. I was at a seeming dead end, stuck in a he said/she said kind of spot.

Or was I?

See, in hunting for Ancient Alien artifacts on the Moon, or anywhere else for that matter, there is always the crucial factor of *context* in evaluating images. If the famous Face on Mars were an isolated landform, it would never have sparked the

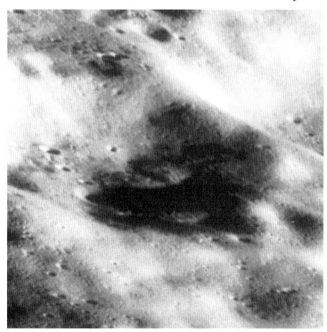

The Crane.

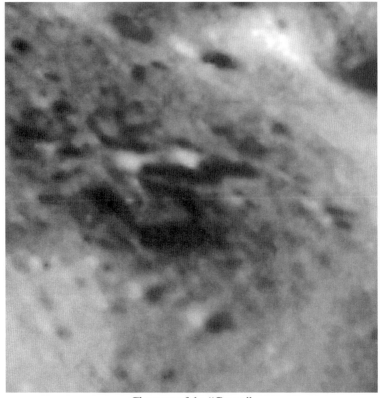

Close-up of the "Crane."

interest it has for over 30 years. The same concept applied here. If the Ziggurat were the only object that was unusual on AS11-38-5564, then we could safely assume that it was a prank or misinformation put out by professionals, and that the rather plain looking official version of the photo was genuine. But if even the sanitized NASA version had other anomalies on it, then the chances that the Ziggurat was real and had been (perhaps hastily) removed from AS11-38-5564 went way up.

So I started looking. And wow. Just wow!

AS11-38-5564 is covered with machinery, structures, buildings, artifacts and Ancient Alien ruins of all types. The hardest task for this book is to cut it down to just the weirdest and most obvious ones.

The first thing to understand about AS11-38-5564 is where the light is coming from. On this image, the sun is coming from the right at an almost 90° angle to the camera. This means that tall objects sticking up in the sunlight will cast shadows to the left at something less than (but pretty close to) 90 degrees to the frame of reference (the camera).

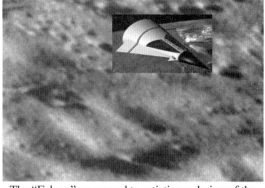

The "Falcon" compared to artistic rendering of the spacecraft from the film *Planet of the Apes*.

The first object I noted was something I decided to call the Crane. It was just sitting there, on a ridge to the west of and not far from the area of the Ziggurat. The crane overlooked a deep, dark hole in the ground which had some very geometric looking "platforms" extending over the opening below.

The crane appeared to have some sort of extension (or cannon) raised and turned toward the right, and the shadow cast by the

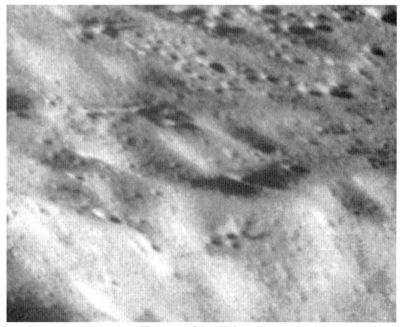

Close-up of the "Falcon."

205

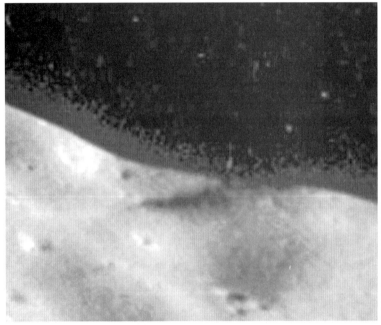

The Daedalus Spire and shadow.

extension and the lower base part of the vehicle were consistent with the lighting. It looked as if somebody had simply parked it at the top of the hill above the opening below.

Not far from the Crane is another object that is sitting atop a ridge and looks distinctly artificial. The "Flacon" looks similar to the spacecraft seen in the original Planet of the Apes movies in the 1960's. It has a pointed nose, a raised main body on the top and two dark, recessed areas that might be the cockpit area. It is pronounced in its symmetry and is sitting atop what might be a hanger or launch platform.

Whatever the Falcon is; a sled, a cable car or some type of launch vehicle, it absolutely does not belong on the Moon in any kind of natural model.

Further down in the image on the edge of one of the largest craters in the photo, is yet another spire. Like its namesake in Sinus Medii, this one is sticking straight up out of the lunar surface and casting a shadow across the sunlit regolith below.

What exactly this particular spire might be is unknown, but again it appears to be made of a semi-translucent material and part of a larger construct around it that is stretching down into the darkness of the crater. At the scale of the photograph, it would have to be several hundred feet tall at least and again completely contradicts the standard models of lunar geology.

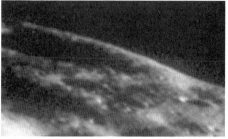

The "gun emplacement." Note recessed area around object.

The Gun Emplacement

The Gun Emplacement is a geometric object consisting of two cylindrical footpads, a central spherical housing (which is casting a shadow on the ground behind it), and a long extension which looks something like a gun or a cannon

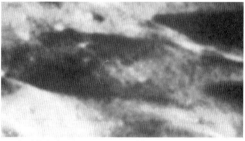

Artificial sub-structure of lunar surface near Daedalus.

extruding from the front. It is emplaced in a recessed bunker or foxhole and sits adjacent to another geometrically square area behind it.

Farther away, there is more strange stuff. If you look closely, you can see extensive areas where the surface appears to be peeled back and an underlying substructure is exposed.

There are numerous areas like this all over AS11-38-5564, but my eye was drawn to the hills above the Ziggurat, where I saw stranger and stranger evidence of Ancient Alien ruins on the

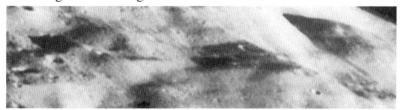

Hillside from AS11-38-5564 showing at least 4 artificial structures/objects.

Moon. The first area had four main features that interested me. They all looked artificial.

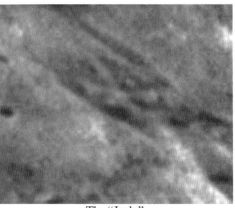

The "Jack."

The first was a distinct cross formation on a hillside near the top of the image. It looked like almost like a flattened jack and had four nodes at the end points. The arms of the jack crossed at 90 degrees to each other. But then things got really weird.

Directly above the "jack" was what looked to be an open hangar door with nothing less than a domed, disk shaped craft parked in it. It appears to have crab-like landing "legs", a couple of evenly spaced dark "vents" or windows of some kind, and a disk shaped body not unlike the Jupiter 2 from *Lost in Space*.

There is also a possible cylindrical attachment of some kind on the left side of the Saucer just above the strut. What its function is I can only guess. There appears to be a great deal of debris in the hangar underneath the saucer, meaning it is likely the area has been abandoned for a long time. But it sure would be fun to land there and look around.

On the hillside right next to the flying saucer hanger is another completely anomalous area that can't be explained by natural processes. Sticking out of the mountainside is a cross hatched pattern of girders and support struts jutting into the air a right angles to the mountainside. It appears to be a series of platforms or terraces constructed to overlook the valley below. What might

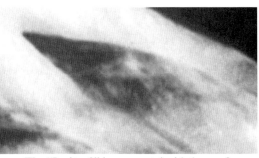

The "Jupiter 2" in a mountainside hanger?

The Beach House.

be beyond the shadows and under the mountain is anybody's guess, but there are several tantalizing clues in the details of this structure. There also look to be pylons below the terraces dug into the mountain to support the dug-in bunker, similar to support pylons at beach houses in places like Malibu, California. The shadows confirm that these platforms are sticking out horizontally, overhanging the mountainside.

As if all this wasn't enough, I saved the weirdest for last.

If you let your eyes keep travelling along this frame from left to right at the same level as the beach house, they will come upon something that simply is too weird to even contemplate, but yet it is there. It's another head.

Sitting on what may be an artificial shelf at roughly the same level as the Saucer and the Malibu beach house is this enormous, human looking head. A metal shelf or platform

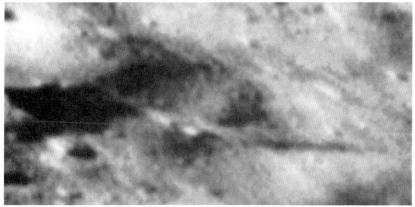

Head shaped object sitting on mountainside next to an artificial platform. Note Stonehenge-like, evenly spaced struts just behind the "Head."

209

The Drill.

extends to the left and a huge portion of it is exposed and casting a shadow down the mountainside. The "head" looks much like "Data's Head" from Shorty crater, but is more on a scale of the statues at Easter Island. Whether it's truly an artifact of some kind or simply an oddly symmetrical boulder with 2 eyes, brow ridges, a nose and what may be mouth is, I suppose, debatable. But after what you've already seen on this photo do you really have any doubts? Also, just behind the "Head" are a series of regularly spaced "struts" emerging from the ground.

The last object I'll make note of, and the one I'm most fascinated with (because it's the most obviously artificial) I call the "Drill."

Sweeping across the terrain and connecting to a mountainside, this long tubular object terminates in a bright white head that looks like something out of a Roto-rooter commercial. The shadow cast by the tube extends from right to left until it is partially buried under some raised ground. The raised ground has a very block like, industrial look to it, and it is probably some kind of maintenance facility for whatever is inside the tube structure. The tube then emerges on the other side and connects to the mountain with what look to be a complex series of tubes and straps. A close-up of the head of the drill shows it is extremely complex—and undeniably artificial.

The only question is what exactly, it's doing there, and what exactly the Ancient Aliens were drilling for.

210

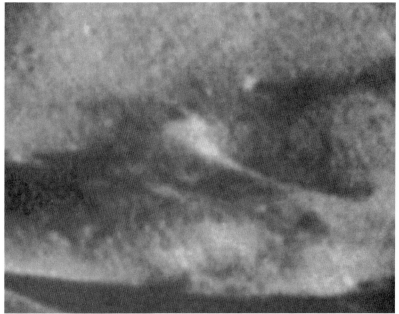

The Drill head. Note multiple attachment points to the mountain.

Which brings us, finally, back to the Ziggurat. I could go on and on showing you more astoundling structures in this and accompanying images, but I think the point is made: There once was a massive, Ancient Alien base on the far side of the Moon near the crater Daedalus. But I still can't get past the Ziggurat. So I started playing with the official NASA version of AS11-38-5564. In looking closely at the side-by-side comparisons, I just couldn't shake the idea that something was wrong. Given everything else on this image, I was now convinced that someone at NASA had fooled around with the image. But how?

Then I saw it. And I understood.

In close-up I couldn't see it except for the V-shaped ramps in front. The walls, the central temple and the dome shaped top simply weren't there. But as I loaded different versions of the image through my windows browser, I began to notice something odd. In the thumbnails of the area, I *could* see the Ziggurat, almost fully formed. Huh?

Then I figured out why this made sense. The human eye, especially when dealing with grayscale images like these, needs

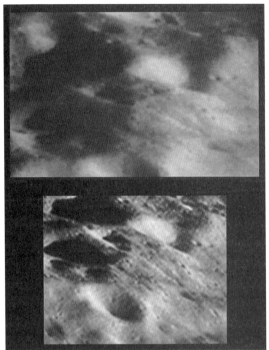

Comparison of the original Ziggurat and contrast
stretched version of NASA data.

contrast to ascertain detail. Without it, we are virtually powerless
to make sense of visual images.

So was that it then? Was the Ziggurat after all just a contrast
artifact of the enhancement process caused by an overzealous
anomaly hunter?

No. No way.

Comparison of the original Ziggurat and contrast stretched
version of NASA data.

For one thing, the original source file just had too much detail.
After working with the "official" version of AS11-38-5564, I had
to reduce it very significantly to get it to look anything at all like
the Ziggurat. Even then, there were several things out of place on
the official version, features in different positions, which led me
to conclude that it had been altered to obscure the details of the
Ziggurat.

What probably happened is that when this image hit the web,
somebody at NASA said "Holy crap! There's a pyramid on this

photo!" and proceeded to add shadows, reduce contrast, introduce noise and generally futz with the picture. This is certainly not unprecedented. Anyone who has read *Dark Mission* knows about the skullduggery that went along with images of the Face on Mars. Satisfied I had found something quite special, I passed the Ziggurat on to my co-author on *Dark Mission*, Richard C. Hoagland. He quickly recognized the importance of the data and went on George Noory's *Coast to Coast AM* program with it on the night of July 20th, 2012 (yeah, that date) and talked about it in the news segment. The reaction from our critics was typical and predictable.

The usual suspects immediately claimed that Hoagland had "hoaxed" the image or that it was a "fraud," and if it wasn't a fraud, then his inability to see it as a fraud was proof he was either a "liar" or "incompetent." The chief purveyor of this nonsense was somebody named Stuart Robbins, on his blog. Robbins has a long history of false and utterly silly accusations against me and Mr. Hoagland, and frequently teams up with someone calling himself "Expat" to attack us within hours of anything we post. "Expat" in fact has made a habit of stalking my radio appearances to ask me in-depth questions along the lines of "are you still beating your wife?"

Now in terms of Mr. Robbins "analysis" of the images, I will simply say that it leaves a lot to be desired. Let me also state that I am no Photoshop expert, like he claims to be, and lack the artistic talent to create anything like Daedalus Ziggurat. I found the image, posted by someone else, period. I enhanced it as best I could and passed it on to Richard for his opinion. I had no idea he would talk about it on *Coast*.

Now I do know enough about image enhancement to know a few things that are relevant. First, because both "as1120pyramid20smallue2.jpg" and the currently posted NASA image are jpegs, they have quality issues and are not truly ideal as research quality documents. In order to do a proper analysis, anyone accusing Mr. Hoagland or myself of fraud would have to obtain a research quality original of AS11-38-5564 and do a high-resolution scan of it under controlled conditions. Neither Mr.

Robbins nor the other self-appointed defenders of true science have done so. In fact, all they are doing is comparing one lossy jpeg document of dubious origin to another. Then, based only on their irrational bias toward NASA and against myself and Mr. Hoagland, they are jumping to the conclusion(s) that we are "frauds" or "incompetents." As I will soon demonstrate, we are neither.

One of the main arguments that Mr. Robbins made on his blog that he cites as proof that we have "drawn in" the Ziggurat is that the image presented by Hoagland has a lot of "noise" in it. Since he was working with Hoagland's enhancement of my enhancement, I guess we can cut him some slack on that. Or not.

The reason there is more noise in the original Ziggurat image is that it was probably scanned from an original and then enlarged, processed, and then reduced for publishing on the web. This is easy to see by the fact that it has a 72 Dots Per Inch resolution, which is standard for the web. This has the effect of making it a bit noisier, but also easier to upload and download from the internet. There is nothing nefarious or questionable about this. In fact, the "Save for Web" tool in Photoshop *automatically* changes the document resolution from say, 300 DPI (the resolution of AS11-38-5564 on the LPI website image) to 72 DPI. This alone will induce noise at deep levels of the image, and contrary to Mr. Robbins assertion, is indicative of nothing except his desire to deceive his readers into thinking there's something unusual about it. Jpeg's are always noisy. It's as simple as that.

One other point here, the Ziggurat, which is miles across in the master image, is way beyond limits of resolution where noise could be a problem. All of the visible features plenty big enough to resolve, even on a lossy jpeg.

Of course, if your intent is to deceive your readers into buying into your own petty biases and jealousy's against people that are more important than you are, you go the extra mile, don't you? What Mr. Robbins didn't tell you is that a large chunk of the "noise" that appears in the image he "processed" was deliberately induced — by him.

On his blog post as he's describing his method of analyzing the image I gave to Hoagland, he lists this little gem:

> For the record, I took the original LPI image and rotated it clockwise 90°. I knew this was the starting point because of the shadows of craters in the image Hoagland presented. After finding the location, I rotated Hoagland's image by 10.96°, and then I *scaled* Hoagland's by 85.28%. [emphasis added]

Now, there is no question that the LPI image contains more total information than "as1120pyramid20smallue2.jpg," which was the source of the enhancement that I sent to Richard. Given that, why would anyone of fair mind, especially a self-appointed expert in Photoshop who claims that "my work over the past twenty years doing image processing and analysis" qualifies him to pass judgment on myself and Mr. Hoagland as "frauds," *reduce* our image by "85.28%?" In fact, anyone who knows anything about image enhancement knows that reducing an image induces *more noise* and reduces detail *by design*. In the image enhancement world, it's known as downsampling. Any competent image enhancement specialist would have enlarged the NASA image instead to bring it in line with the size of the original. This would have the effect of

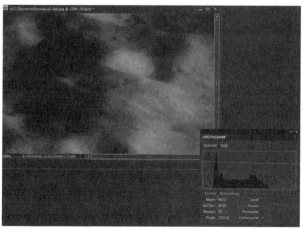

Histogram showing the dynamic visual range of the original Daedalus Ziggurat image, "as1120pyramid20smallue2.jpg"

actually making the NASA image *better*, rather than making the original enhancement *worse*.

Given that he claims that we "deliberately added noise and reduced the quality" elsewhere in his blog, this would seem to be kind of a big mistake, wouldn't it Mr. Robbins? Unless of course, it wasn't a mistake at all. Unless it was a deliberate act of deception foisted upon his readers.

His use of the word "scaled" is a further indictment of his intentions. He was hoping no one would notice that he had degenerated the data, rather than enhanced it. The word "reduced" would have been far more honest. But I already knew I wasn't dealing with an honest critic anyway...

And of course, as usual, the gang that can't shoot straight had made another huge error, one that proves that it is the official NASA image, and not ours, that has been manipulated and faked. An error that anyone who claims to have even the most rudimentary knowledge about image enhancement would never make.

You see, Photoshop has a tool called a histogram. What a histogram does is analyze the dynamic range of a given image (or highlighted section of an image). This can act like a digital finger print to help us determine if an image or a part of an image has been manipulated changed, altered or enhanced. In a grayscale image such as "as1120pyramid20smallue2.jpg," or the official AS11-38-5564 image from the NASA/LPI website, the image has a possible range of up to 256 shades of gray. In reading the histogram, absolute black has a value of zero (0), and occupies the spot on the graph at the far left. Pure, bright white would have a value of 256 and occupy the spot on the far right of the histogram graph. When we apply the histogram to the original image, "as1120pyramid20smallue2. jpg," we can see that it has a fairly wide dynamic range, from color 24 on the left of the graph (meaning something less than black) to color 184 on the right, which is something less than pure white. What this means is that the image has a dynamic range of 160 colors, or shades of gray, and the colors at the extreme dark and light ranges of the spectrum have been lost somewhere along the way. This is not ideal, but it does mean only that my original image has probably lost some shading information at some point.

The NASA image however, is a different story. It contains the full range of 256 shades of gray, which would seem to mean it is a "better" image and more likely to be the authentic article. But the histogram contains something curious.

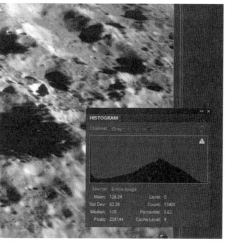

Close-up of histogram. Note huge spike at color level zero–Absolute black.

There, at the far left of the graph, is shade zero, or absolute black. What is curious is that by far the most pixels in the NASA image are absolute black. In fact, color zero and the next few colors over (near absolute black) make up more than 33% of the entire image. What that means is that somebody put a lot of black and near black in the NASA image.

Now the truth is, almost nothing in any image is ever absolute black or white. But to find that (by far) the biggest number of pixels is absolute, perfect black is more than a little suspicious. On a properly processed image, the histogram should be pretty much a bell curve, bulging in the middle and then dropping off at both ends. The NASA image doesn't do this. It spikes on shade zero, absolute black.

So what's the most likely reason for this? I can think of only one. Somebody took a paintbrush tool, set it to color zero, or absolute black, and went to town on it.

And I can prove it.

As you look at the Ziggurat image, obviously some of the most tell-tale signs of artificiality are in the shadowed areas, particularly the shadow cast by the west wall. In my enhanced version of the original, you can plainly see the wall and how it casts a shadow into the depression below.

Contrast enhanced image of the Daedalus Ziggurat.

But on the NASA version, this shadow is so dark that there is virtually no detail there at all. No wall, no shades of gray in the crater, just pitch-black darkness. Everything you need to see to confirm the wall and the artificiality of the Ziggurat is simply not there. It's gone.

Adobe, the makers of Photoshop, are crystal clear on the meaning of this: "If many pixels are bunched up at either the shadow or highlight ends of the chart, it may indicate that image detail in the shadows or highlights may be clipped—blocked up as pure black or pure white. There is little you can do to recover this type of image."

In other words, it's a deliberate manipulation of the image in question.

Unenhanced image of the NASA version of the
Daedalus Ziggurat.

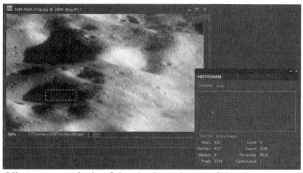

Histogram analysis of the shadowed area of the west wall of the Ziggurat.

What the histograms show us is that while the image produced by NASA has a wide dynamic range, the areas of shadow, where the details that make the Ziggurat stand out reside, have virtually no dynamic range. They're absolute black. And that can only mean one thing; they were painted over by someone at NASA with a black paintbrush tool.

To test this, all we have to do is select the areas in question, and examine their histograms. The results are conclusive.

In the shadowed area where we can see the wall in the original version, virtually *all* of the pixels are absolute, indisputably black. The same applies to the dark shadowed area just behind the temple section of the Ziggurat. These specific areas of the image—the ones that would provide the smoking gun for the Ziggurat's artificiality—have been "blocked up as pure black" by NASA.

Case closed.

Oh, I'm aware that the deceptive Mr. Robbins has tried to cover his ass by claiming that

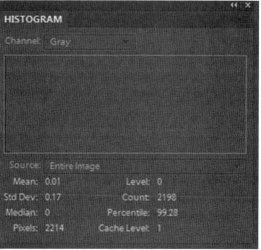

Histogram analysis of the shadowed area of the west wall of the Ziggurat showing 99% of pixels in area are shade zero; absolute black.

219

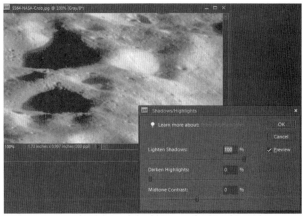

Enhancement of NASA original image of AS11-38-5564 showing areas of the Daedalus Ziggurat digitally altered by NASA.

because of the lighting conditions on the Moon some areas are
absolute black. But the fact is that contrary to his fallacious claims
that there are "no crater wall nor mountain to scatter light onto
it," there is a very bright, sunlit area just to left of the selection
marquee that should be scattering plenty of light into the shadowed
area, but isn't.

Because NASA blacked it out.

Still don't believe me? OK, here's one more test. I'll show you
exactly where the brush strokes are…

You see, in Photoshop, there is another enhancement tool that's
commonly used in cases like this. It's one that somebody with
"20 years" of experience doing Photoshop should know about, but
Mr. Robbins apparently doesn't. It's called the "Adjust Lighting…
Shadows/Highlights" tool. It contains a nifty little slider named
"Lighten Shadows:" that allows you to, well, lighten shadows.
Let's see what it does to the shadows on NASA's image.

Well golly, look at that. It allows us to see exactly where the
goons at NASA applied the pure black paintbrush to the shadows
around the Ziggurat. You can even see that they probably used an
airbrush tool, because the edges are jagged.

One final point, if you look at the shadow created by the
western wall of the Ziggurat that has been blacked out by the NASA
paintbrush artist, you can see that he made a mistake and only

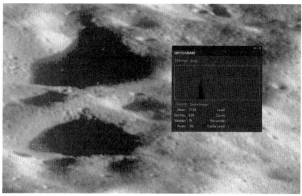

Histogram analysis of western side of the Daedalus Ziggurat
shadow not blacked out by NASA.

covered the offending parts and not the complete shadow. Some
of the area appears gray, and if we zoom up and scan it we can see
that it contains a moderate range of gray shades. What this means
is that Robbins assertion about there being "no light scattering"
in the area is also proven wrong. Absolutely, completely wrong.
It's patently obvious to any competent Photoshop user that the
original lighting in this area was almost certainly very close to the
lighting seen in "as1120pyramid20smallue2.jpg," which means
that his claims that the lighting on that image is "not possible" and
therefore a sign of "fraud" is also complete idiocy.

So as it stands, while we may never know the exact origins
of the original Daedalus Ziggurat image, we know with absolute
certainty that the NASA version of AS-11-38-5564 has been
deliberately faked to hide something from us. My bet is it's a really
big Ziggurat on the Moon.

What I have proven here is that the chief critics of the Daedalus
Ziggurat are not only wrong, they are foolishly and sloppily wrong.
Either they didn't want to study the images in the depth that I
did before releasing it to Mr. Hoagland, or they simply lacked the
intellectual capacity to do so. I lean towards the latter, but let's
face it, buffoons like Mr. Robbins and Expat are just haters who
attack everything I do, no matter how many times I embarrass
them. I doubt this will be the last time I have to respond to these
louts, but I certainly hope it's now obvious they don't know what

they're talking about. I would say they were liars and/or idiots, but that's their territory. The difference is I can actually prove what I say.

At any rate, we now have indisputable proof that NASA alters and puts fraudulent images on the web. They managed, through a combination of the techniques I have exposed here, to obscure the reality of Daedalus Ziggurat, at least to some degree. But they have in the process exposed themselves once again as frauds and deceivers, and brought down some of their staunchest allies in the process. I only wish Dr. Phil Plait had chimed in on it too.

Fortunately for NASA, there will be plenty more "useful idiots" to take the place of those that have now been exposed. Fortunately for us, the guy at NASA who did the fake missed about a dozen other Ancient Alien artifacts on the photo. So regardless of whether we ever get the real, unaltered version of the Ziggurat to compare

AS17-151-23260.

to the original, we at least have those artifacts to point to.

ULO'S

Before we conclude, I would be remiss if I didn't mention the work of one especially talented Ancient Alien artifacts hunter. In 2008, photographer Allan Sturm came out with an e-book entitled "ULO's, Unidentified Lunar Objects Revealed in NASA Photography." It contains what I consider some of the best images of bases, ruins and industrial complexes all over the lunar surface, and from a wide variety of missions. Allan's skill in colorizing the features is also a tremendous asset to the book, which helps give the readers the crucial visual cues to the structures they're looking at. I wanted Allan's work to be a central part of this book, but for personal reasons he declined. However, I consider some of his findings so crucial (and so obvious) that I don't feel they can be excluded from this book.

Some of my favorite "ULO's" can be found on NASA frames AS08-13-2267, AS15-M-2502, and especially AS17-151-23260. If you can find a copy of his now out of print book, I encourage you to get one. If not, go get copies of these images and start looking around for yourself. You'll be stunned at what you find.

But I want to focus especially on AS17-151-23260. It's a wide shot of the crater Copernicus, and that is an area of the Moon that has interested me for some time. The photo shows not only the crater, but the very uneven ground on the ejecta blanket surrounding it.

The area of greatest interest is just beyond the crater rim on the left side of the photo. It doesn't take much enhancement work to see that the entire area is so uneven because the ejecta material is

Sectional enlargement of Copernicus crater rim from AS17-151-23260.

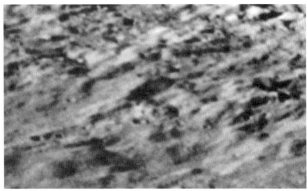

AS17-151-23260.

covering the ruins of a vast complex of artificial structures buried just beneath the lunar sands. There are all manner of partially buried mechanical and industrial structures there, easily revealed by their geometric shape and organized layout.

Closer inspection reveals what look like roads, over passes, long tube structures and shattered, battered equipment still hanging above gaping holes in the surface.

But the most impressively artificial structure is an area I call the "Power Plant."

The power plant reminds me of the rebel base on ice planet Hoth from the Star Wars movie "The Empire Strikes Back" because there are no less than 6 partially buried wheel-like "generators" with regularly placed spokes in them. To the left of the close-up is a long tube (which extends well beyond this cropped sectional) which connects to some other part of the installation. To the right is another gaping hole in the lunar floor that has girders and other construction materials hanging over it.

Once again, there can be no doubt, debate or rational argument that these structures are anything but artificial. There is zero chance that the wheels or anything else in the immediate area was formed

Close-up of the Power Plant from AS17-151-23260.

any kind of natural process. This, along with everything else you've seen in this book is de facto *proof* of the Ancient Alien presence on the Moon.

But there is still just a bit more.

In 2009, NASA launched the LCROSS (Lunar Crater Observation and Sensing Satellite) to the Moon in an effort to detect water in a crater there. Launched together with

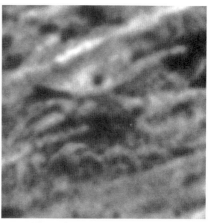

Ultra close-up of the "wheels."

the Lunar Reconnaissance Orbiter, LCROSS detached from the orbiter and its Centaur upper stage was directed to a collision course with a south polar crater named Caebeus. The idea was that what was left of LCROSS (called the Shepherding Spacecraft) would study the plume generated from the impact (equivalent to 2 tons of TNT) by flying through it. In this way, NASA hoped to determine if there was water in sufficient amounts there to one supply a lunar base. Later, the Shepherding Spacecraft would itself crash into the Moon.

Expectations were so high at NASA that there would be a visible plume of material for visual and spectral analysis that they encouraged amateur astronomers to make observations of they saw. There was just one problem; when the Centaur stage hit, there wasn't any plume at all. Instead, the booster just disappeared into the lunar interior like a rock dropping into a tar pit. Photos of the area taken by major telescopes (like Mount Palomar) showed no

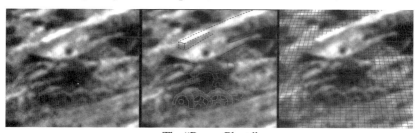

The "Power Plant."

225

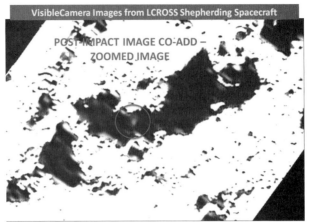

NASA enhancement of Visible Camera Image from
LCROSS Shepherding Spacecraft.

evidence of the impact at all.

So what had happened?

Well, as usual, my co-author on *Dark Mission*, Richard C. Hoagland, was right on top of it. Seven days later, NASA claimed that the Centaur impact (which was supposed to be the big one) had generated a feeble few "bright pixels" on the Shepherding Spacecraft's visual light camera. They released an image showing what they purported to be the "plume."

Instead, what it showed was the Centaur booster had crashed into an area that was covered in geometric shapes, and had apparently failed to generate a plume because it had crashed through into a vast complex below the lunar surface. It seemed obvious from the images that what NASA was claiming as the "plume" was actually a reflection off of some kind of tubing or

Enhancement showing geometric substructure beneath the lunar
surface in the crater Caebeus.

226

structure under the surface. So in attempting to find water, NASA had instead inadvertently "bombed" an Ancient Alien underground base.

Which should just about wrap it up as far as the doubters are concerned.

---

[1] http://www.disclose.tv/forum/pictures-of-pyramids-on-the-moon-t53252-20.html

EPILOGUE

Through the course of these pages, we have seen what I consider to be indisputable proof that there was once an Ancient Alien presence on the Moon on a vast scale. We've looked at the early histories of lunar anomalies, from the inexplicable bright flashes to the odd color changes and "vaporous" appearance of these transient events. We have come up with a theory to account for these early observations, and then had that theory—that these bright reflections are intense specular reflections off a titanic glass meteor shield constructed eons ago—confirmed. We've seen the otherwise inexplicable Surveyor 6 image of the Sun's light bent around the lunar surface exactly as an atmosphere would. We've seen the Apollo 15 Trans Lunar Injection photo showing the mysterious "airglow" layer around the Moon—a layer of glass. We've seen the Tower and the Shard and the Crisium Spire, all indisputable evidence that this glass shield exists, even if it has been battered and beaten almost to dust in certain places. And we've seen the ground truth; the domes over the Cleomedes craters, the odd structures on the rim of Tycho, the astonishing ruins surrounding Copernicus crater, and the stunning reality of the Daedalus Ziggurat. But even with all of this, we are left with a few final questions; who built these structures? And more importantly, what happened to them?

Up until now, our search for Ancient Aliens on the Moon has been like a police procedural TV show, gathering evidence, making suppositions, looking for supporting evidence. But now, we must delve into the realm of the purely speculative.

The question of who built them would seem to be an easy one to answer. If we believe in the history books that teach us that we went from dumb old cavemen with stone tools and bearskins to smart, modern us with our hydrogen bombs and our striped toothpaste in one straight line, then they *must* be alien ruins. They must have come from somewhere else. Whether it was the giant

229

Nefilim of the bible, the bug-eyed Grays of the modern motif or the proverbial little green men, they weren't human. Not in any way. Yet, so much of what we see on the Moon looks *familiar*. It looks like stuff that we would build. The architecture has a *human* feel to it. How then to explain that?

If the Ancient Alien theory implies anything, it is that perhaps we did not evolve in that simple straight line from caveman to quarterback. That there were some ups and downs in between.

In my previous books, *Dark Mission* and *The Choice*, I pointed out the substantial and credible evidence to support this idea, so I won't recite all of it here. But I'm drawn again to the Hopi prophecies I discussed in *The Choice*. The Hopi tell us that this is the 4ᵗʰ world of Man, and that the previous 3 worlds were destroyed in unimaginable catastrophes of their own making. In the last World they say, our intellects had become so vast that we had many technological inventions not seen yet in this World. Inventions that would have enabled us to build bases on the Moon, and presumably elsewhere in the solar system. Maybe we even built a race of sentient robots, *Cylons* if you will—to do the dangerous work on the Moon and to make our lives easier. Maybe that's what "Data's Head" is.

I think that is some of what we're seeing on these images. We're seeing the last remnants of that previous version of Man. We're seeing the bases and habitats and industries and machines that they left behind when they were overtaken by whatever disaster struck them down.

Yet clearly, some of these ruins are older than that. What kind of technology could re-shape a mountain in to a hexagonal structure made of titanium and glass, and towering miles above the Taurus-Littrow valley? What kind of technology can plow and modify the landscape around Daedalus and Copernicus, the way *somebody* did so long ago? Those kinds of efforts seem beyond the simple technologies we can imagine. They seem almost god-like.

So I think in the end we're seeing a bit of both. Some of what we've found on the Moon is Alien, built by a civilization that has no relationship to human beings. But some of it—a lot of it in fact—was built by somebody else. Somebody a bit more closely related to us. Maybe that's what Neil Armstrong meant when he said "That's one small step for Man... one giant leap for Mankind." Maybe he didn't make a mistake in that speech at all. Maybe what he was heralding was the return of a genetically related "Mankind" to their proper place amongst the stars.

As to what happened to the builders, be they humanoid or Ancient Alien, it looks to me like some of what is left was simply abandoned. Whoever was doing whatever they were doing on our Moon finished their project and left. What they left behind has been slowly decaying ever since. Other parts of these ruins look more haphazard, like they were suddenly and decisively overwhelmed by the cataclysm the Hopi talk about. Whatever the reasons, it seems clear that what we've found in the pages of this book is very ancient, and very dead.

And those facts would also help explain why NASA has been less than curious and far from cooperative in pursuing this line of thinking. The Brooking Report told them to keep quiet if they found evidence of ET ruins around the solar system. But what if they not only found that, but found evidence that these godlike ET's got their asses kicked by some unimaginable catastrophe? That, it seems to me, would be reason enough to suppress (but not completely hide) what they really found on the Moon. After all, there is always that first, uncomfortable question at the proverbial press conference if you announce that Aliens once roamed the Moon:

"Excuse me, what happened to them?"

In my next book, *Ancient Aliens on Mars*, I will examine this possible solar system wide catastrophe in greater detail, right down to the timing of it. But for now, what seems more important is to assume that the Moon holds many mysteries, as we've now seen, many surprises, as we now know, and perhaps, many answers, as we now hope.

So it's time to go back. And this time, tell the truth about what we find.

Mike Bara is a *New York Times* Bestselling author and lecturer. He began his writing career after spending more than 25 years as an engineering consultant for major aerospace companies, where he was a card-carrying member of the Military/Industrial complex. A self-described "Born Again conspiracy theorist," Mike's first book *Dark Mission-The Secret History of NASA* (co-authored with the venerable Richard C. Hoagland) was a New York Times bestseller in 2007. *The Choice* from New Page Books was published in 2010.

Mike has made numerous public appearances lecturing on the subjects of space science, NASA, physics and the link between science and spirit, and has been a featured guest on radio programs like Coast to Coast AM with George Noory. He is a regular contributor to the History Channel programs "Ancient Aliens" and "America's Book of Secrets," both of which are now showing on the H2 channel.

THE FANTASTIC INVENTIONS OF NIKOLA TESLA
by Nikola Tesla with David Hatcher Childress
This book is a readable compendium of patents, diagrams, photos and explanations of the many incredible inventions of the originator of the modern era of electrification. In Tesla's own words are such topics as wireless transmission of power, death rays, and radio-controlled airships. In addition, rare material on a secret city built at a remote jungle site in South America by one of Tesla's students, Guglielmo Marconi. Marconi's secret group claims to have built flying saucers in the 1940s and to have gone to Mars in the early 1950s! Incredible photos of these Tesla craft are included. •His plan to transmit free electricity into the atmosphere. •How electrical devices would work using only small antennas. •Why unlimited power could be utilized anywhere on earth. •How radio and radar technology can be used as death-ray weapons in Star Wars.
342 PAGES. 6x9 PAPERBACK. ILLUSTRATED. $16.95. CODE: FINT

PRODIGAL GENIUS
The Life of Nikola Tesla
by John J. O'Neill
This special edition of O'Neill's book has many rare photographs of Tesla and his most advanced inventions. Tesla's eccentric personality gives his life story a strange romantic quality. He made his first million before he was forty, yet gave up his royalties in a gesture of friendship, and died almost in poverty. Tesla could see an invention in 3-D, from every angle, within his mind, before it was built; how he refused to accept the Nobel Prize; his friendships with Mark Twain, George Westinghouse and competition with Thomas Edison. Tesla is revealed as a figure of genius whose influence on the world reaches into the far future. Deluxe, illustrated edition.
408 pages. 6x9 Paperback. Illustrated. Bibliography. $18.95. Code: PRG

TAPPING THE ZERO POINT ENERGY
Free Energy & Anti-Gravity in Today's Physics
by Moray B. King
King explains how free energy and anti-gravity are possible. The theories of the zero point energy maintain there are tremendous fluctuations of electrical field energy imbedded within the fabric of space. This book tells how, in the 1930s, inventor T. Henry Moray could produce a fifty kilowatt "free energy" machine; how an electrified plasma vortex creates anti-gravity; how the Pons/Fleischmann "cold fusion" experiment could produce tremendous heat without fusion; and how certain experiments might produce a gravitational anomaly.
180 PAGES. 5x8 PAPERBACK. ILLUSTRATED. $12.95. CODE: TAP

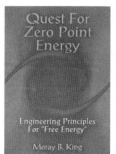

QUEST FOR ZERO-POINT ENERGY
Engineering Principles for "Free Energy"
by Moray B. King
King expands, with diagrams, on how free energy and anti-gravity are possible. The theories of zero point energy maintain there are tremendous fluctuations of electrical field energy embedded within the fabric of space. King explains the following topics: TFundamentals of a Zero-Point Energy Technology; Vacuum Energy Vortices; The Super Tube; Charge Clusters: The Basis of Zero-Point Energy Inventions; Vortex Filaments, Torsion Fields and the Zero-Point Energy; Transforming the Planet with a Zero-Point Energy Experiment; Dual Vortex Forms: The Key to a Large Zero-Point Energy Coherence. Packed with diagrams, patents and photos.
224 PAGES. 6x9 PAPERBACK. ILLUSTRATED. $14.95. CODE: QZPE

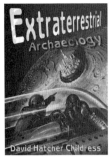

EXTRATERRESTRIAL ARCHEOLOGY
Edited by David Hatcher Childress

Using official NASA and Soviet photos, this book seeks to prove that many of the planets (and moons) of our solar system are in some way inhabited by intelligent life. NASA photos of pyramids and domed cities on the moon; Pyramids and giant statues on Mars; Hollow Moons of Mars and other Planets; Robot Mining Vehicles that move about the Moon processing valuable metals; A British Scientist who discovered a tunnel on the Moon, and other "bottomless craters;" Structural Anomalies on Venus, Saturn, Jupiter, Mercury,Uranus & Neptune; Plus more. Highly illustrated with photos, diagrams and maps!

320 Pages. 8x11 Paperback. Illustrated. $19.95. Code: ETA

INVISIBLE RESIDENTS
The Reality of Underwater UFOS
By Ivan T. Sanderson, Foreword by David Hatcher Childress

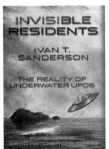

Sanderson puts forward the theory that "OINTS"—Other Intelligences—live under the Earth's oceans. Sanderson postulates that the OINTS are behind many UFO sightings as well as the mysterious disappearances of aircraft and ships in the Bermuda Triangle. What better place to have an impenetrable base than deep within the oceans of the planet? Yet, if UFOs are coming from beneath our oceans or lakes, does it necessarily mean that there is another civilization besides our own that is responsible?

298 Pages. 6x9 Paperback. Illustrated. $16.95. Code: INVS

WEATHER WARFARE
The Military's Plan to Draft Mother Nature
by Jerry E. Smith

Weather modification in the form of cloud seeding to increase snow packs in the Sierras or suppress hail over Kansas is now an everyday affair. Underground nuclear tests in Nevada have set off earthquakes. A Russian company has been offering to sell typhoons (hurricanes) on demand since the 1990s. Scientists have been searching for ways to move hurricanes for over fifty years. In the same amount of time we went from the Wright Brothers to Neil Armstrong. Hundreds of environmental and weather modifying technologies have been patented in the United States alone – and hundreds more are being developed in civilian, academic, military and quasi-military laboratories around the world *at this moment!* Numerous ongoing military programs do inject aerosols at high altitude for communications and surveillance operations.

304 Pages. 6x9 Paperback. Illustrated. Bib. $18.95. Code: WWAR

HAARP
The Ultimate Weapon of the Conspiracy
by Jerry Smith

The HAARP project in Alaska is one of the most controversial projects ever undertaken by the U.S. Government. Jerry Smith gives us the history of the HAARP project and explains how works, in technically correct yet easy to understand language. At at worst, HAARP could be the most dangerous device ever created, a futuristic technology that is everything from super-beam weapon to world-wide mind control device. Topics include Over-the-Horizon Radar and HAARP, Mind Control, ELF and HAARP, The Telsa Connection, The Russian Woodpecker, GWEN & HAARP, Earth Penetrating Tomography, Weather Modification, Secret Science of the Conspiracy, more. Includes the complete 1987 Eastlund patent for his pulsed super-weapon that he claims was stolen by the HAARP Project.

256 pages. 6x9 Paperback. Illustrated. Bib. $14.95. Code: HARP

ATLANTIS & THE POWER SYSTEM OF THE GODS
by David Hatcher Childress and Bill Clendenon
Childress' fascinating analysis of Nikola Tesla's broadcast system in light of Edgar Cayce's "Terrible Crystal" and the obelisks of ancient Egypt and Ethiopia. Includes: Atlantis and its crystal power towers that broadcast energy; how these incredible power stations may still exist today; inventor Nikola Tesla's nearly identical system of power transmission; Mercury Proton Gyros and mercury vortex propulsion; more. Richly illustrated, and packed with evidence that Atlantis not only existed—it had a world-wide energy system more sophisticated than ours today.
246 PAGES. 6x9 PAPERBACK. ILLUSTRATED. $15.95. CODE: APSG

TECHNOLOGY OF THE GODS
The Incredible Sciences of the Ancients
by David Hatcher Childress

Childress looks at the technology that was allegedly used in Atlantis and the theory that the Great Pyramid of Egypt was originally a gigantic power station. He examines tales of ancient flight and the technology that it involved; how the ancients used electricity; megalithic building techniques; the use of crystal lenses and the fire from the gods; evidence of various high tech weapons in the past, including atomic weapons; ancient metallurgy and heavy machinery; the role of modern inventors such as Nikola Tesla in bringing ancient technology back into modern use; impossible artifacts; and more.
356 PAGES. 6x9 PAPERBACK. ILLUSTRATED. BIBLIOGRAPHY. $16.95. CODE: TGOD

VIMANA AIRCRAFT OF ANCIENT INDIA & ATLANTIS
by David Hatcher Childress, introduction by Ivan T. Sanderson

In this incredible volume on ancient India, authentic Indian texts such as the *Ramayana* and the *Mahabharata* are used to prove that ancient aircraft were in use more than four thousand years ago. Included in this book is the entire Fourth Century BC manuscript *Vimaanika Shastra* by the ancient author Maharishi Bharadwaaja. Also included are chapters on Atlantean technology, the incredible Rama Empire of India and the devastating wars that destroyed it.
334 PAGES. 6x9 PAPERBACK. ILLUSTRATED. $15.95. CODE: VAA

GRAVITATIONAL MANIPULATION OF DOMED CRAFT
UFO Propulsion Dynamics
by Paul E. Potter

Potter's precise and lavish illustrations allow the reader to enter directly into the realm of the advanced technological engineer and to understand, quite straightforwardly, the aliens' methods of energy manipulation: their methods of electrical power generation; how they purposely designed their craft to employ the kinds of energy dynamics that are exclusive to space (discoverable in our astrophysics) in order that their craft may generate both attractive and repulsive gravitational forces; their control over the mass-density matrix surrounding their craft enabling them to alter their physical dimensions and even manufacture their own frame of reference in respect to time. Includes a 16-page color insert.
624 pages. 7x10 Paperback. Illustrated. References. $24.00. Code: GMDC

THE TESLA PAPERS
Nikola Tesla on Free Energy & Wireless Transmission of Power
by Nikola Tesla, edited by David Hatcher Childress

David Hatcher Childress takes us into the incredible world of Nikola Tesla and his amazing inventions. Tesla's fantastic vision of the future, including wireless power, anti-gravity, free energy and highly advanced solar power. Also included are some of the papers, patents and material collected on Tesla at the Colorado Springs Tesla Symposiums, including papers on: •The Secret History of Wireless Transmission •Tesla and the Magnifying Transmitter •Design and Construction of a Half-Wave Tesla Coil •Electrostatics: A Key to Free Energy •Progress in Zero-Point Energy Research •Electromagnetic Energy from Antennas to Atoms •Tesla's Particle Beam Technology •Fundamental Excitatory Modes of the Earth-Ionosphere Cavity

325 PAGES. 8X10 PAPERBACK. ILLUSTRATED. $16.95. CODE: TTP

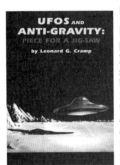

UFOS AND ANTI-GRAVITY
Piece For A Jig-Saw
by Leonard G. Cramp

Leonard G. Cramp's 1966 classic book on flying saucer propulsion and suppressed technology is a highly technical look at the UFO phenomena by a trained scientist. Cramp first introduces the idea of 'anti-gravity' and introduces us to the various theories of gravitation. He then examines the technology necessary to build a flying saucer and examines in great detail the technical aspects of such a craft. Cramp's book is a wealth of material and diagrams on flying saucers, anti-gravity, suppressed technology, G-fields and UFOs. Chapters include Crossroads of Aerodymanics, Aerodynamic Saucers, Limitations of Rocketry, Gravitation and the Ether, Gravitational Spaceships, G-Field Lift Effects, The Bi-Field Theory, VTOL and Hovercraft, Analysis of UFO photos, more.

388 PAGES. 6x9 PAPERBACK. ILLUSTRATED. $16.95. CODE: UAG

THE COSMIC MATRIX
Piece for a Jig-Saw, Part Two
by Leonard G. Cramp

Cramp examines anti-gravity effects and theorizes that this super-science used by the craft—described in detail in the book—can lift mankind into a new level of technology, transportation and understanding of the universe. The book takes a close look at gravity control, time travel, and the interlocking web of energy between all planets in our solar system with Leonard's unique technical diagrams. A fantastic voyage into the present and future!

364 PAGES. 6x9 PAPERBACK. ILLUSTRATED. BIBLIOGRAPHY. $16.00. CODE: CMX

THE A.T. FACTOR
A Scientists Encounter with UFOs
by Leonard Cramp

British aerospace engineer Cramp began much of the scientific anti-gravity and UFO propulsion analysis back in 1955 with his landmark book *Space, Gravity & the Flying Saucer* (out-of-print and rare). In this final book, Cramp brings to a close his detailed and controversial study of UFOs and Anti-Gravity.

324 PAGES. 6x9 PAPERBACK. ILLUSTRATED. BIBLIOGRAPHY. INDEX. $16.95. CODE: ATF

THE TIME TRAVEL HANDBOOK
edited by David Hatcher Childress
The Time Travel Handbook takes the reader beyond the government experiments and deep into the uncharted territory of early time travellers such as Nikola Tesla and Guglielmo Marconi and their alleged time travel experiments, as well as the Wilson Brothers of EMI and their connection to the Philadelphia Experiment—the U.S. Navy's forays into invisibility, time travel, and teleportation. Childress looks into the claims of time travelling individuals, and investigates the unusual claim that the pyramids on Mars were built in the future and sent back in time. A highly visual, large format book, with patents, photos and schematics.
316 PAGES. 7x10 PAPERBACK. ILLUSTRATED. $16.95. CODE: TTH

DARK MOON
Apollo and the Whistleblowers
by Mary Bennett and David Percy
Did you know a second craft was going to the Moon at the same time as Apollo 11? Do you know that potentially lethal radiation is prevalent throughout deep space? Do you know there are serious discrepancies in the account of the Apollo 13 'accident'? Did you know that 'live' color TV from the Moon was not actually live at all? Did you know that the Lunar Surface Camera had no viewfinder? Do you know that lighting was used in the Apollo photographs—yet no lighting equipment was taken to the Moon? All these questions, and more, are discussed in great detail by British researchers Bennett and Percy in *Dark Moon*, the definitive book (nearly 600 pages) on the possible faking of the Apollo Moon missions.
568 PAGES. 6x9 PAPERBACK. ILLUSTRATED. BIBLIOGRAPHY. INDEX. $32.00. CODE: DMO

THE CRYSTAL SKULLS
Astonishing Portals to Man's Past
by David Hatcher Childress and Stephen S. Mehler
Childress introduces the technology and lore of crystals, and then plunges into the turbulent times of the Mexican Revolution form the backdrop for the rollicking adventures of Ambrose Bierce, the renowned journalist who went missing in the jungles in 1913, and F.A. Mitchell-Hedges, the notorious adventurer who emerged from the jungles with the most famous of the crystal skulls. Mehler shares his extensive knowledge of and experience with crystal skulls. Having been involved in the field since the 1980s, he has personally examined many of the most influential skulls, and has worked with the leaders in crystal skull research, including the inimitable Nick Nocerino, who developed a meticulous methodology for the purpose of examining the skulls.
294 pages. 6x9 Paperback. Illustrated. Bibliography. $18.95. Code: CRSK

SECRETS OF THE MYSTERIOUS VALLEY
by Christopher O'Brien
No other region in North America features the variety and intensity of unusual phenomena found in the world's largest alpine valley, the San Luis Valley of Colorado and New Mexico. Since 1989, Christopher O'Brien has documented thousands of high-strange accounts that report UFOs, ghosts, crypto-creatures, cattle mutilations, skinwalkers and sorcerers, along with portal areas, secret underground bases and covert military activity. This mysterious region at the top of North America has a higher incidence of UFO reports than any other area of the continent and is the publicized birthplace of the "cattle mutilation" mystery. Hundreds of animals have been found strangely slain during waves of anomalous aerial craft sightings. Is the government directly involved? Are there underground bases here? Does the military fly exotic aerial craft in this valley that are radar-invisible below 18,000 feet?
460 PAGES. 6x9 PAPERBACK. ILLUSTRATED. BIBLIOGRAPHY. $19.95. CODE: SOMV

SECRETS OF THE UNIFIED FIELD
The Philadelphia Experiment, the Nazi Bell, and the Discarded Theory
by Joseph P. Farrell

Farrell examines the discarded Unified Field Theory. American and German wartime scientists determined that, while the theory was incomplete, it could nevertheless be engineered. Chapters include: The Meanings of "Torsion"; The Mistake in Unified Field Theories and Their Discarding by Contemporary Physics; Three Routes to the Doomsday Weapon: Quantum Potential, Torsion, and Vortices; Tesla's Meeting with FDR; Arnold Sommerfeld and Electromagnetic Radar Stealth; Electromagnetic Phase Conjugations, Phase Conjugate Mirrors, and Templates; The Unified Field Theory, the Torsion Tensor, and Igor Witkowski's Idea of the Plasma Focus; tons more.

340 pages. 6x9 Paperback. Illustrated. Bibliography. Index. $18.95. Code: SOUF

THE GIZA DEATH STAR
The Paleophysics of the Great Pyramid & the Military Complex at Giza
by Joseph P. Farrell

Was the Giza complex part of a military installation over 10,000 years ago? Chapters include: An Archaeology of Mass Destruction, Thoth and Theories; The Machine Hypothesis; Pythagoras, Plato, Planck, and the Pyramid; The Weapon Hypothesis; Encoded Harmonics of the Planck Units in the Great Pyramid; High Freqquency Direct Current "Impulse" Technology; The Grand Gallery and its Crystals: Gravito-acoustic Resonators; The Other Two Large Pyramids; the "Causeways," and the "Temples"; A Phase Conjugate Howitzer; Evidence of the Use of Weapons of Mass Destruction in Ancient Times; more.

290 PAGES. 6x9 PAPERBACK. ILLUSTRATED. $16.95. CODE: GDS

THE GIZA DEATH STAR DEPLOYED
The Physics & Engineering of the Great Pyramid
by Joseph P. Farrell

Farrell expands on his thesis that the Great Pyramid was a maser, designed as a weapon and eventually deployed—with disastrous results to the solar system. Includes: Exploding Planets: A Brief History of the Exoteric and Esoteric Investigations of the Great Pyramid; No Machines, Please!; The Stargate Conspiracy; The Scalar Weapons; Message or Machine?; A Tesla Analysis of the Putative Physics and Engineering of the Giza Death Star; Cohering the Zero Point, Vacuum Energy, Flux: Feedback Loops and Tetrahedral Physics; and more.

290 PAGES. 6x9 PAPERBACK. ILLUSTRATED. $16.95. CODE: GDSD

THE GIZA DEATH STAR DESTROYED
The Ancient War For Future Science
by Joseph P. Farrell

Farrell moves on to events of the final days of the Giza Death Star and its awesome power. These final events, eventually leading up to the destruction of this giant machine, are dissected one by one, leading us to the eventual abandonment of the Giza Military Complex—an event that hurled civilization back into the Stone Age. Chapters include: The Mars-Earth Connection; The Lost "Root Races" and the Moral Reasons for the Flood; The Destruction of Krypton: The Electrodynamic Solar System, Exploding Planets and Ancient Wars; Turning the Stream of the Flood: the Origin of Secret Societies and Esoteric Traditions; The Quest to Recover Ancient Mega-Technology; Non-Equilibrium Paleophysics; Monatomic Paleophysics; Frequencies, Vortices and Mass Particles; "Acoustic" Intensity of Fields; The Pyramid of Crystals; tons more.

292 pages. 6x9 paperback. Illustrated. $16.95. Code: GDES

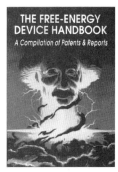

THE FREE-ENERGY DEVICE HANDBOOK
A Compilation of Patents and Reports
by David Hatcher Childress

A large-format compilation of various patents, papers, descriptions and diagrams concerning free-energy devices and systems. *The Free-Energy Device Handbook* is a visual tool for experimenters and researchers into magnetic motors and other "over-unity" devices. With chapters on the Adams Motor, the Hans Coler Generator, cold fusion, superconductors, "N" machines, space-energy generators, Nikola Tesla, T. Townsend Brown, and the latest in free-energy devices. Packed with photos, technical diagrams, patents and fascinating information, this book belongs on every science shelf.
292 PAGES. 8x10 PAPERBACK. ILLUSTRATED. $16.95. CODE: FEH

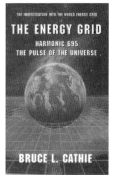

THE ENERGY GRID
Harmonic 695, The Pulse of the Universe
by Captain Bruce Cathie

This is the breakthrough book that explores the incredible potential of the Energy Grid and the Earth's Unified Field all around us. Cathie's first book, *Harmonic 33*, was published in 1968 when he was a commercial pilot in New Zealand. Since then, Captain Bruce Cathie has been the premier investigator into the amazing potential of the infinite energy that surrounds our planet every microsecond. Cathie investigates the Harmonics of Light and how the Energy Grid is created. In this amazing book are chapters on UFO Propulsion, Nikola Tesla, Unified Equations, the Mysterious Aerials, Pythagoras & the Grid, Nuclear Detonation and the Grid, Maps of the Ancients, an Australian Stonehenge examined, more.
255 PAGES. 6x9 TRADEPAPER. ILLUSTRATED. $15.95. CODE: TEG

THE BRIDGE TO INFINITY
Harmonic 371244
by Captain Bruce Cathie

Cathie has popularized the concept that the earth is crisscrossed by an electromagnetic grid system that can be used for anti-gravity, free energy, levitation and more. The book includes a new analysis of the harmonic nature of reality, acoustic levitation, pyramid power, harmonic receiver towers and UFO propulsion. It concludes that today's scientists have at their command a fantastic store of knowledge with which to advance the welfare of the human race.
204 PAGES. 6x9 TRADEPAPER. ILLUSTRATED. $14.95. CODE: BTF

THE HARMONIC CONQUEST OF SPACE
by Captain Bruce Cathie

Chapters include: Mathematics of the World Grid; the Harmonics of Hiroshima and Nagasaki; Harmonic Transmission and Receiving; the Link Between Human Brain Waves; the Cavity Resonance between the Earth; the Ionosphere and Gravity; Edgar Cayce—the Harmonics of the Subconscious; Stonehenge; the Harmonics of the Moon; the Pyramids of Mars; Nikola Tesla's Electric Car; the Robert Adams Pulsed Electric Motor Generator; Harmonic Clues to the Unified Field; and more. Also included are tables showing the harmonic relations between the earth's magnetic field, the speed of light, and anti-gravity/gravity acceleration at different points on the earth's surface. New chapters in this edition on the giant stone spheres of Costa Rica, Atomic Tests and Volcanic Activity, and a chapter on Ayers Rock analysed with Stone Mountain, Georgia.
248 PAGES. 6x9. PAPERBACK. ILLUSTRATED. BIBLIOGRAPHY. $16.95. CODE: HCS

ORDER FORM

10% Discount When You Order 3 or More Items!

One Adventure Place
P.O. Box 74
Kempton, Illinois 60946
United States of America
Tel.: 815-253-6390 • Fax: 815-253-6300
Email: auphq@frontiernet.net
http://www.adventuresunlimitedpress.com

ORDERING INSTRUCTIONS

✓ Remit by USD$ Check, Money Order or Credit Card

✓ Visa, Master Card, Discover & AmEx Accepted

✓ Paypal Payments Can Be Made To:

    info@wexclub.com

✓ Prices May Change Without Notice

✓ 10% Discount for 3 or More Items

SHIPPING CHARGES

United States

✓ Postal Book Rate { $4.50 First Item / 50¢ Each Additional Item

✓ POSTAL BOOK RATE Cannot Be Tracked!
    Not responsible for non-delivery.

✓ Priority Mail { $6.00 First Item / $2.00 Each Additional Item

✓ UPS { $7.00 First Item / $1.50 Each Additional Item

    NOTE: UPS Delivery Available to Mainland USA Only

Canada

✓ Postal Air Mail { $15.00 First Item / $2.50 Each Additional Item

✓ Personal Checks or Bank Drafts MUST BE

    US$ and Drawn on a US Bank

✓ Canadian Postal Money Orders OK

✓ Payment MUST BE US$

All Other Countries

✓ Sorry, No Surface Delivery!

✓ Postal Air Mail { $19.00 First Item / $6.00 Each Additional Item

✓ Checks and Money Orders MUST BE US$
    and Drawn on a US Bank or branch.

✓ Paypal Payments Can Be Made in US$ To:
    info@wexclub.com

SPECIAL NOTES

✓ RETAILERS: Standard Discounts Available

✓ BACKORDERS: We Backorder all Out-of-
Stock Items Unless Otherwise Requested

✓ PRO FORMA INVOICES: Available on Request

✓ DVD Return Policy: Replace defective DVDs only

ORDER ONLINE AT: www.adventuresunlimitedpress.com

**10% Discount When You Order
3 or More Items!**

Please check: ✓

☐ This is my first order   ☐ I have ordered before

Name

Address

City

State/Province   Postal Code

Country

Phone: Day   Evening

Fax   Email

Item Code	Item Description	Qty	Total

Please check: ✓   Subtotal ▶

Less Discount-10% for 3 or more items ▶

☐ Postal-Surface   Balance ▶

☐ Postal-Air Mail   Illinois Residents 6.25% Sales Tax ▶
   (Priority in USA)   Previous Credit ▶

☐ UPS   Shipping ▶
   (Mainland USA only) Total (check/MO in USD$ only) ▶

☐ Visa/MasterCard/Discover/American Express

Card Number:

Expiration Date: Security Code:

✓ SEND A CATALOG TO A FRIEND: